Ethics of Science and Technology Assessment
Volume 31

Book Series of the Europäische Akademie zur Erforschung
von Folgen wissenschaftlich-technischer Entwicklungen
Bad Neuenahr-Ahrweiler GmbH
edited by Carl Friedrich Gethmann

G. Hanekamp (Ed.)

Business Ethics of Innovation

Springer

Series Editor
Professor Dr. Dr. h.c. Carl Friedrich Gethmann
Europäische Akademie GmbH
Wilhelmstr. 56, 53474 Bad Neuenahr-Ahrweiler, Germany

Editor
Dr. Gerd Hanekamp
Wissenschaftsrat
Brohler Str. 11, 50968 Köln, Germany

Desk Editor
Friederike Wütscher
Europäische Akademie GmbH
Wilhelmstr. 56, 53474 Bad Neuenahr-Ahrweiler, Germany

ISSN 1860-4803
ISBN 978-3-540-72309-7 Springer Berlin Heidelberg New York

This work is subject to copyright. All rights are reserved, whether the whole or part of the material is concerned, specifically the rights of translation, reprinting, reuse of illustrations, recitation, broadcasting, reproduction on microfilm or in any other way, and storage in data banks. Duplication of this publication or parts thereof is permitted only under the provisions of the German Copyright Law of September 9, 1965, in its current version, and permission for use must always be obtained from Springer. Violations are liable for prosecution under the German Copyright Law.

Springer is a part of Springer Science+Business Media
springer.com

© Springer-Verlag Berlin Heidelberg 2007

The use of general descriptive names, registered names, trademarks, etc. in this publication does not imply, even in the absence of a specific statement, that such names are exempt from the relevant protective laws and regulations and therefore free for general use.

Typesetting: Köllen-Druck+Verlag GmbH, Bonn + Berlin
Production: LE-TEX Jelonek, Schmidt & Vöckler GbR, Leipzig
Cover: eStudioCalamar S.L., F.Steinen-Broo, Girona, Spanien

Printed on acid-free paper   60/3180/YL - 5 4 3 2 1 0

zur Erforschung von Folgen wissenschaftlich-technischer Entwicklungen
Bad Neuenahr-Ahrweiler GmbH

Direktor: Professor Dr. Dr. h. c. Carl Friedrich Gethmann

The Europäische Akademie

The *Europäische Akademie zur Erforschung von Folgen wissenschaftlich-technischer Entwicklungen GmbH* is concerned with the scientific study of consequences of scientific and technological advance for the individual and social life and for the natural environment. The Europäische Akademie intends to contribute to a rational way of society of dealing with the consequences of scientific and technological developments. This aim is mainly realised in the development of recommendations for options to act, from the point of view of long-term societal acceptance. The work of the Europäische Akademie mostly takes place in temporary interdisciplinary project groups, whose members are recognised scientists from European universities. Overarching issues, e.g. from the fields of Technology Assessment or Ethic of Science, are dealt with by the staff of the Europäische Akademie.

The Series

The series Ethics of Science and Technology Assessment *(Wissenschaftsethik und Technikfolgenbeurteilung)* serves to publish the results of the work of the Europäische Akademie. It is published by the academy's director. Besides the final results of the project groups the series includes volumes on general questions of ethics of science and technology assessment as well as other monographic studies.

Preface

The Europäische Akademie zur Erforschung von Folgen wissenschaftlich-technischer Entwicklungen Bad Neuenahr-Ahrweiler GmbH is concerned with the scientific study of the consequences of scientific and technological advance for individual and societal life and for the natural environment. The main focus is the examination of mid- and long-term processes that are especially influenced by natural- and engineering sciences and the medical disciplines. As an independent scientific institution, the Europäische Akademie pursues a dialogue with politics and society.

The work of the Europäische Akademie is mainly conducted by temporary interdisciplinary project teams that develop recommendations for a long-term reliable science and technology policy. Above that, the Europäische Akademie organises international conferences on timely issues that are connected to its research programme.

The present volume takes an applied ethics perspective on a group of phenomena that are related to the concept of innovation. The Europäische Akademie has already devoted some of her work to research in this field and is pursuing this effort. Hopefully the endeavour started with the conference *Business Ethics of Innovation* documented in this volume will help to appropriately assess innovations not only academically but also and consecutively in those contexts where actors have to base their decisions on pertinent reflections.

Bad Neuenahr-Ahrweiler, December 2006   Carl Friedrich Gethmann

Foreword

Presently consequences of research in the sciences and in the arts as well as those of technological advances are reflected in several fields of academic activity: Technology assessment, history of science, applied ethics, many fields of the social sciences to name only a few. Between these interdisciplinary endeavours only few linkages exist.

An example is the research on innovations. Bio-medical innovations are discussed in technology assessment, bioethics, business ethics, economics and business administration. For innovations in the IT sector bioethics needs to be replaced by activities concerned with access to knowledge and digital technology.

With the 2005 fall conference that took place 11–13 September 2005 at Engers Castle, Neuwied, the Europäische Akademie GmbH intended to develop a scheme for the integration of the disciplines involved in these fields. Since business ethics is the practical discipline concerned with economic activity it was chosen to be the point of departure. The results are encouraging and are documented in this volume.

I would like to thank the staff of the Europäische Akademie GmbH for their support, notably Sevim Kiliç and Margret Pauels for their organisational ingenuity as well as Friederike Wütscher, who prepared this volume with charming vigour. The academy's director Carl Friedrich Gethmann I thank for his support to realise the conference and this volume.

Further acknowledgements go to the Land Rheinland-Pfalz, the German Aerospace Center (DLR) and the Schering AG for their generous grants.

Bad Neuenahr-Ahrweiler and Gerd Hanekamp
Cologne, December 2006

List of Authors

Crane, Andrew, Professor of Policy and holds the George R. Gardiner Chair in Business Ethics at the Schulich School of Business, York University, Toronto. He is interested in various aspects of business ethics, including the role of morality in marketing and consumption; the contribution of evolutionary narratives to environmental management; the implementation of fair trade policies; and the contribution of Foucauldian thought to business ethics. Recent work appeared in "Academy of Management Review", "Journal of Business Research", "Organization Studies", "Journal of Business Ethics" and "Business Ethics Quarterly". He holds a BSc from Warwick University and a PhD in Business Studies from Nottingham University. Previously, he was a Professor of Business Ethics and Director of the MBA in CSR at Nottingham University Business School, UK.
E-mail: acrane@schulich.yorku.ca

Eaton, Margaret L., Pharm.D., J.D. Lecturer in Management, Graduate School of Business, Stanford University (Faculty Affiliate: Center for Social Innovation). Past positions: Stanford Center for Biomedical Ethics, Stanford University Office of the General Counsel, and Lecturer, Stanford University School of Medicine. Studied law at University of California at San Francisco. Studied pharmaceutics at Union University and Duquesne University and served as Associate Professor, University of Minnesota School of Pharmacy.
E-mail: maggie@ronandmaggie.com

Hanekamp, Gerd, Dr. phil. Dipl.-Chem., studied chemistry at Heidelberg and Marburg and at the École Nationale de Chimie in Lille; 1996 Doctorate in philosophy at the University of Marburg; from 1996–2003 junior/senior scientist, from 2003–2005 deputy director, Europäische Akademie Bad Neuenahr-Ahrweiler GmbH, presently on leave for a position at the Office of the Science Council; fields of research: applied ethics, philosophy of science, theory of the social sciences, technology assessment.
E-mail: hanekamp@wissenschaftsrat.de

Matten, Dirk, Professor of Policy, holds the Hewlett-Packard Chair in Corporate Social Responsibility at the Schulich School of Business, York University, Toronto. He is interested in political and international aspects of ethics and CSR, recent work appeared in "Academy of Management Review", "Jour-

nal of Management Studies", "Organization Studies", "Journal of Business Ethics" and "Business Ethics Quarterly". His first degree in business is from the University of Essen (Dipl.-Kfm.) and his PhD and Habilitation from Heinrich-Heine-University Düsseldorf. In 2004–2006 he was Professor of Business Ethics and Director of the Centre for Research into Sustainability at Royal Holloway, University of London, UK.
E-mail: dmatten@schulich.yorku.ca

Moon, Jeremy, BA, PhD, Professor of Corporate Social Responsibility and Director of the International Centre for Corporate Social Responsibility (ICCSR) at Nottingham University Business School. His previous appointments include Chair of the Political Science Department at the University of Western Australia. He held a number of visiting positions at universities around the globe, including Princeton, McGill, and Cambridge, and was a Jean Monnet Research Fellow at the European University Institute in Florence. He holds a BA and a PhD in Political Science from Exeter University. Jeremy has published widely in the areas of Government and CSR; CSR in Europe; CSR and Globalisation and Conceptualising and Theorising CSR.
E-mail: jeremy.moon@nottingham.ac.uk

Nüttgens, Markus, Professor Dr., studied Business Administration at the University of Saarland 1984–1989 (Dipl.Kfm.) and 1995 (Dr. rer. oek.). 1989–2001 Scientific Assistant at Institute of Information Systems (IWi), 2001–2003 Deputy Professor of Information Systems at the University Trier, since 2004 Full Professor of Information Systems at the University Hamburg. Main interests: Information Systems Management, Information- and Business Process Management, Business Standard Software, Open Source/Access, IT-Entrepreneurship.
E-mail: markus.nuettgens@wiso.uni-hamburg.de

Seiter, Andreas, Dr. med., studied medicine at Friedrich-Alexander-Universität in Erlangen, Germany 1974–1980; intern in aneasthesiology and internal medicine at Gunzenhausen hospital 1982–84; 1984–2003 various management positions in medical operations, marketing and external relations at Sandoz AG in Germany and Switzerland, later merged with Ciba to Novartis AG; 2004–2006 Pharmaceutical Fellow at the World Bank; since May 2006 Senior Health Specialist, Pharmaceuticals, The World Bank.
E-mail: aseiter@worldbank.org

Selgelid, Michael J., received a BSEng in Biomedical Engineering at Duke University and an MA and PhD in Philosophy at the University of California, San Diego. His doctoral dissertation examined ethical issues associated with eugenics. He previously worked as a Postdoctoral Research Fellow and Lecturer, respectively, in the Division of Bioethics and Philosophy Department at the University of the Witwatersrand in Johannesburg, South Africa; and as Sesquicentenary Lecturer in Bioethics in the Unit for History and Philosophy of Science and the Centre for Values, Ethics and the Law in Medicine (VELIM) at the University of Sydney. He is currently a Senior Research Fellow at the Centre for Applied Philosophy and Public Ethics (CAPPE) and the Menzies Centre for Health Policy at The Australian National University in Canberra. His research over the past five years has focused on the history of and ethical issues associated with infectious disease. He recently completed co-editing a book titled "Ethics and Infectious Disease" (Blackwell, 2006).

Steinmann, Horst, Professor em. Dr. Dres. h.c., studied Business Administration at the Universität Göttingen 1954–1959 (Dipl.-Kfm.) and at the Institut Européen d'Administration des Affaires (INSEAD), Fontainebleau, 1964/65 (MBA). 1962 Dr. rer. nat. and 1967 Habilitation (TU Clausthal). 1968–1970 Full Professor of Operations Research at the FU Berlin; 1970–1999 Full Professor of Business Administration and General Management at the Universität Erlangen-Nürnberg. Doctor honoris causa Universität Bern (1996) and University Robert Schuman Strassbourg (1999). Founder of the German Business Ethics Network (1993). Main interests: General Management, Business Ethics, Corporate Governance, Philosophy of Science.
E-mail: horst.steinmann@wiso.uni- erlangen.de

Table of Contents

Preface

Foreword

List of Authors

Business Ethics of Innovation. An Introduction
Gerd Hanekamp 1

Corporate Ethics and Globalization. Global Rules and Private Actors
Horst Steinmann 7

Research Priorities, Profits, and Public Goods:
The Case of Drug Resistant Disease
Michael J. Selgelid 27

Ethical Issues Associated with Pharmaceutical Innovation
Margaret L. Eaton 39

Corporate Responsibility for Innovation – A Citizenship Framework
Dirk Matten, Andy Crane and Jeremy Moon................................... 63

Access to Medicines and the Innovation Dilemma –
Can Pharmaceutical Multinationals be Good Corporate Citizens?
Andreas Seiter 89

IT Innovations & Open Source: A Question of Business Ethics or
Business Model?
Markus Nüttgens 101

Business Ethics of Innovation. An Introduction

Gerd Hanekamp

Innovations are said to be the key drivers of economic development. They are the root for competitive advantages that allow firms to take the lead in particular markets (Albach 1994, Freeman and Soete 1997). Seen as the establishment of new products – that correspond to new means for specific ends – on markets they are a genuine object of economic and business research (Brockhoff 1999). In a widened view they are social phenomena that every now and then substantially change aspects of everyday life. The development of antibiotic medications as well as the introduction of personal computing are of this kind. Nanotechnologically facilitated drug delivery, converging technologies and pervasive computing might be contemporary candidates. These innovations are a societal challenge and as such are meanwhile attributed a considerable share of attention in policy-making.

The changing structures of political decision-making described as governance (Benz 2004) as well as rising awareness of consumers in terms of the impact new products may have are a challenge for actors in the business field. Firms are involved in governance processes. They are expected to contribute to the development of the framework they act in. Furthermore, firms are supposed to not only provide new means for specific ends but to contribute to the orientational task of specifying new ends and reflecting upon the consequences that are connected to their attainment. They have to think about the availability of their products especially if the cure from a fatal disease or social inclusion with respect to a new technology depend on them. This orientational task is a critical endeavor in the Kantian sense and deserves further attention as a philosophical project.

A 'Business Ethics of Innovation' is situated at the intersection between various fields of applied ethics, political philosophy, social philosophy, technology assessment, economics, business administration and sociology.

Objectives

Business ethics of innovation explores the philosophical aspects and implications of innovations from the perspective of the firm (Hanekamp 2005). This endeavor most likely will have consequences for this perspective. It will result in a critical assessment of how firms act with respect to innovations. Particularly the results will offer routes to an integration of this critical reflection on innovations and their development in the decision making process of the firm itself.

A business ethics of innovation is driven by two heuristic schemes. One describes the perspectives on and the influences of innovations as well as the social entities that are involved. The other describes aspects of decision-making and conflict solving. These models have to be enriched and modified as the relevant bodies of knowledge are considered and integrated.

There are different perspectives on innovations that can be illustrated as follows:

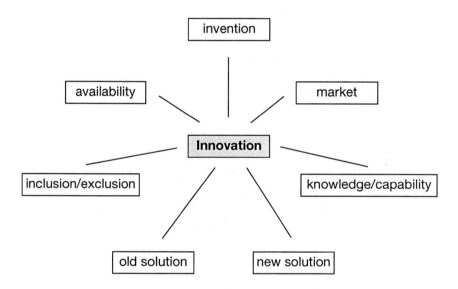

The usual assessment of innovations in business is connected with the key words invention, market, old and new solution. With the latter two the terminology of Joseph Schumpeter (1942) is taken up. Pragmatically these refer to a particular relation of means to ends. A new solution can correspond to a new means for established ends or to means to reach new ends. The consideration of other aspects such as the availability of the new solution, the inclusion or exclusion from its use or the relationship to knowledge and capabilities is not part of standard procedures in spite of pertinent discussions in the literature.

Innovations are of interest not only for firms but likewise and partly for different reasons for the state, for society, for different communities and the economy. The dimension of aspects is thus supplemented by a dimension of actors:

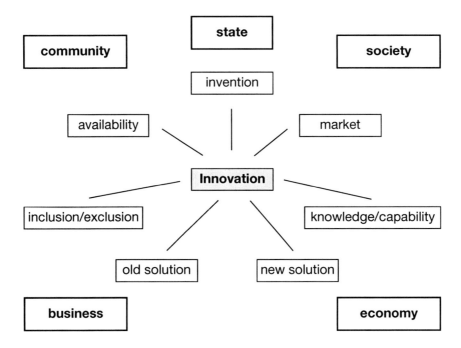

In the case of HIV/AIDS medication e.g. there are the states where the pharmaceutical firms producing the medication are operating, the developing world states that are most intensely affected, there are different communities in these states possibly having a particular view on the disease, there are the firms themselves, and there are the respective societies and economies as more abstract entities that are affected and have an influence on the situation. All these can view innovations in terms of the aspects listed above. For a business actor it can be essential to consider them comprehensively in order to adequately address the corresponding interests or problems in their decisions.

For a particular problem the two dimensions can be systematically presented in a simple two-dimensional matrix (see below). The cells of the

	business	state	economy	society	community
Market					
new solution					
old solution					
invention					
knowledge/ capabilities					
exlusion/ inclusion					
availability					

matrix traditionally corresponding to the core interest of firms are marked in grey. It is an objective of a business ethics of innovation to widen this perspective in the two dimensions of the above matrix and to integrate it into a specific approach to business ethics (arrows in the matrix above).

For the business actor the vertical extension means additional aspects to consider, the horizontal extension means other perspectives that are considered for the decision or conflict at stake.

The decision or conflict that needs to be taken or resolved can be systematized in a step-wise procedure. Such a procedure can be developed starting with the following heuristic scheme (Hanekamp 2004, 2006):

0. *"View of affairs":* Which view of the world, human beings and oneself is presupposed?
1. *Stake, conflict, objectives:* What is it about?
2. *Urgency:* In which time frame does the matter have to be settled?
3. *Stakeholder:* Who is concerned? Who has to be involved?
4. *Rules and presuppositions:* Which rules are relevant, which presuppositions are affected?
5. *Availability:* Which types of changes are possible or acceptable respectively? Which part of the framework is available for change? Which types of changes is one entitled to make?
6. *Functionality:* Which performance of organizations and institutions are concerned and possibly altered?
7. *Legitimacy:* Which basis of legitimacy can the relevant institutions rely on?
8. Balance and Decision

The 'view of affairs' is underlying the whole scheme and refers to the philosophical discussion of an anthropological foundation of ethics. The 'view of affairs' can change, e.g. within the discussion of availabilities in step 5. A balanced decision always includes a critical attitude towards these views, whether it be to confirm or to revise.

The scheme is supposed to structure the handling of decisions and conflicts reconstructively. Steps 4, 5, 6 and 7 underline the importance of a detailed presuppositional analysis. Furthermore, with step 7 the importance of institutions is stressed – that is not given proper consideration in many approaches of ethical reflection.

Step 8 laconically makes reference to many traits of the discussions in ethics and social philosophy. Importantly, the view of affairs will come into play here. Of course, an ethical approach cannot free the decision maker from deciding, by e.g. following a calculus-type decision scheme. Ethics can only structure decision-making and conflict-solving and point out presuppositions and what follows from them.

In order to implement the above procedure within a business ethics of innovation the interface between philosophy and business in terms of the widened view on innovations has to be developed. In a first step the management, especially the organizational literature has to be assessed. This assessment will be useful for the analysis of how business is dealing with innovations and their different aspects as well as for the refinement of the procedure in terms of its expediency for the business environment. It will thus be of major importance for the organizational implementation of the above procedure in the business process. The availability of a straight forward organizational structure is key for the successful establishment of the procedure. Social and political philosophy as well as the available work on governance has to complement this endeavor. The widened view on innovation thus has its counterpart in the view on management and organization.

The need for this kind of normative approach to business administration, especially for organization has been recognized (Margolis and Walsh 2003). The agenda put forward there is not thoroughly philosophical. The analytical starting point, however, is shared and can be applied analogously in other areas: In business administration research the tension of economic and non-economic objectives is generally tried to be resolved by either empirically or theoretically *reconciling* them by proving that the non-economic objectives are instrumental for the achievement of the economic ones.

This compares to an analytic predominance of economic thinking and prevents researchers from exploring the full scope of options and methodologically leads to problems of the 'self-fulfilling-prophecy'-type. A business ethics of innovation can be a cure for this shortcoming. Its widened perspective, however, means a considerable widening of research fields involved.

References and further reading

Albach H (ed) (1994) Culture and Technical Innovation. A cross-cultural analysis and policy recommendations. Berlin
Beauchamp TL, Childress JF (2001) Principles of Biomedical Ethics. New York
Beauchamp TL, Bowie NE (2004) Ethical Theory and Business. Prentice Hall, Upper Saddle River
Benz A (ed) (2004) Governance – Regieren in komplexen Regelsystemen. VS, Wiesbaden
Brockhoff K (1999) Forschung und Entwicklung. München
Chadwick RF (ed.) (2002) Applied Ethics. Critical Concepts in Philosophy, 6 Bde. London
Crane A, Matten D (2004) Business Ethics. Oxford
De George RT (2003) The Ethics of Information Technology and Business. Oxford
De George RT (2005) Business Ethics. Upper Saddle River
Eaton ML (2004) Ethics and the Business of Bioscience. Stanford
Farmer P (1999) Infections and Inequalities. The Modern Plagues. Berkeley
Freeman C, Soete L (1997) The Economics of Industrial Innovation. Cambridge
Goodin RE (1982) Political Theory and Public Policy. Chicago
Goodin RE (2005) Reflective Democracy. Oxford
Günther K (1988) Der Sinn für Angemessenheit. Anwendungsdiskurse in Moral und Recht. Frankfurt a.M.
Hanekamp G (2004) Kulturalistische Unternehmensethik. Begründung und Anwendung. In: Friesen H, Berr K (Hrsg) Angewandte Ethik im Spannungsfeld von Begründung und Anwendung. Frankfurt a.M.
Hanekamp G (2005) Business Ethics of Innovation. Poiesis & Praxis 3:310–314, DOI 10.1007/s10202-005-0001-4
Hanekamp G (2006) Kulturalistische Unternehmensethik. Ein Stufenschema für Entscheidungen und Konfliktlösungen. In: Beschorner T, Schmitt M, Unternehmerische Verantwortung in Zeiten kulturellen Wandels. Hampp
Löhr A (2002) Macht und betriebswirtschaftliche Organisationstheorie. Rekonstruktion und Neuorientierung. Habilitationsschrift, Wirtschafts- und Sozialwissenschaftliche Fakultät der Friedrich-Alexander-Universität Erlangen-Nürnberg
Meyer-Krahmer F, Lange S (ed) (1999) Geisteswissenschaften und Innovationen. Heidelberg
Ortmann G (2004) Als ob. Fiktionen und Organisationen. VS, Wiesbaden
Schumpeter JA (1942) Capitalism, Socialism and Democracy. New York
Steinmann H (2004) Begründungsprobleme einer Unternehmensethik – insbesondere das „Anfangsproblem". Die Unternehmung 02/2004
Velasquez MG (2002) Business Ethics. Concepts and Cases. Prentice Hall, Upper Saddle River

Corporate Ethics and Globalization – Global Rules and Private Actors

Horst Steinmann

1 Introduction

> The fundamental defect of global society today is not that the reach of corporations is too big, but that our ability to govern is too small. We face governance gaps and governance failures on a monumental scale. Our core challenge, therefore, is to stimulate social and political processes that will help bridge the gaps and reduce the failures. The dynamic interplay between business, civil society, and the public sector constitutes an essential platform from which to mount the campaign.

It was John Ruggie from Harvard Law School, appointed by Kofi Annan as "Special Representative on the issue of human rights and transnational corporations", who made this statement on the occasion of the Carl Bertelsmann Prize International Symposium in 2002 (Ruggie 2002:2). As a professor of "International Affairs" Ruggie supported the UN-Secretary in developing and introducing the "Global Compact". The Global Compact is an attempt to involve corporations and other institutions in a world-wide program to respect and support – within their sphere of influence – 10 fundamental principles, concerning *human rights, social standards (labor and working conditions), the environment and freedom from corruption.*

It is this political context of global governance and the involvement of multinational corporations in processes of creating new political and social institutions which form the actual background of this paper. What I would like to elaborate on a bit is the role which *corporate ethics* could and indeed should play in this context (see also Steinmann 2003). This is a highly controversial topic in the discipline of management theory because what is at stake here is nothing less than the rationale, the "raison d'être", of the private corporation in capitalist market economies of the future. But the topic is also highly controversial in and between other academic disciplines, like economics, politics, and here especially the "Theory of International Relations", moreover: sociology, philosophy, political philosophy, international law, and so forth.

In view of this state of affairs and in line with central philosophical ideas of the German philosophical school of "Methodolocigal Constructivism" (Lorenzen 1987, Janich 2001) I regard it to be an important task of management theory (as the first and immediate addressee of business practice) to take on, where necessary, an *interactive role* as a kind of *mediator* and *translator* between problems of business practice and other academic disciplines able to contribute to the solution of the problem at hand (Steinmann and Scherer 2002); and this with the objective to come, first of all, to a common understanding of the *empirical situation* within which to act and, secondly, working on adequate *normative* orientations for change. In order to fulfill this role management theory has to grasp and to conceptualize, on the level of language, the respective problems of business practice; and this in such a way that the basic notions required for tackling these problems can be picked up by other disciplines for further (interdisciplinary) research.

It is from this methodological perspective that we have proposed, within the *national* context, a notion of corporate ethics (Steinmann and Löhr 1994). I shall present this notion very briefly (2). Its core idea is that managers of *private* corporations should be held responsible and accountable not only for making sufficient profits but for contributing, at the same time, directly to the *public* interest by peacefully solving those conflicts with its stakeholders which arise out of corporate strategy. The notion of "strategy-centered peaceful conflict resolution" is at the heart of my concept of corporate ethics, and this not merely as a *laudable activity* (as is the case e.g. with philanthropic corporate givings), but as a *moral duty* which is not at the discreation of management.

I want to show that this concept is so general that it can be applied to the *global* political context mentioned by Ruggie (3). To do this I shall sketch what Kofie Annan in his millenium address (Annan 2000) has called "Global Public Policy Networks" as a representative example of what is actually going on in practice. I maintain that what these networks offer could be understood as a manifestation of corporate ethics.

What follows, then, are some remarks on a number of urgent questions for future research of different disciplines needed to develop a theory of the firm which can cope with the problems of corporate governance in a globalized world (4). What I am not going to do is to outline such a theory; its time is still to come. I shall, instead, mention at the end some basic ideas which may have the potential to improve the actual situation (5).

2 A Notion of Corporate Ethics

(1) Let me start with a quote from an article which appeared in the prestigious "Economist", January 22, 2005 issue, entitled "The good company" (Crook 2005). This article has aroused wide critical discussions and comments within the field of business ethics.

According to the author Clive Crook it is a pity that the movement of *corporate social responsibility* (CSR) has won the battle of ideas. Let me just remind you that this idea has gained great political support by the European Union, as an important basis for the sustainable economic development of

Europe; "CSR-Europe" is an organization which has got quite a lot of money from the EU to further clarify and develop the idea, but has had no great impact on the practical situation as yet. Crook rightly criticizes the muddled thinking on CSR which means a range of fairly different things to the union of concerned executives. He blames "good corporate citizens" for virtually having surrendered to the belief that capitalism is wicked. He highlights that the contribution of profit to the public good is misunderstood by the protagonists of the idea. And he asks for an ethics of business which puts two, and only two, *constraints* on the proper business goal of maximizing owner value, namely (1) respect for property rights and (2) distributive justice in the sense of pay linked to performance and promotion on merits.

I share the view of the author that the actual discussion about corporate social responsibility, corporate citizenship, corporate ethics and similar labels is still quite confusing because clear notions are lacking (Matten and Crane 2005, Scherer and Baumann 2004). I do believe, however, that the rather conservative idea of "business ethics" put forward by Crook fails to meet the relevant practical normative requirements of todays corporate management.

(2) My understanding of corporate ethics starts with the simple insight that, for a society of free and equal individuals, good governance, in order to successfully *coordinate* human actions as the focal point of the argument, requires to reconcile *individual freedom* with the *unity of society*. In our context of *corporate* governance this would demand that corporate ethics should *directly* link the raison d'être of the private corporation to the public interest as follows:

- Drawing on the work of Paul Lorenzen (1987:233) I propose to look at the *end* to make *peace* more stable in and between nations (or societies) as a *reasonable* general expression of what is usually called the "public interest" or "common good". Peaceful resolution of conflicts should be regarded as the central value and *normative* basis of corporate ethics; and, secondly,
- that corporations should contribute to the public interest directly by peacefully solving those conflicts with corporate stakeholders which follow or might follow from their profit oriented corporate strategies. In other words: corporate ethics is about the *means* by which corporations try to make profits. Are these means acceptable from a moral point of view? Think of money laundering, corruption and violation of broadly accepted social or environmental standards as only three examples of moral scandals well known from reports in the media.

(3) These two ideas underlying my concept of corporate ethics may help to improve the *factual* legitimation of corporate governance. But before I elaborate a bit more on it let me just point out that corporate ethics might also contribute to corporate legitimacy on a more *formal* (legal) basis, namely by

integrating it into (national) company law, as part of director's duties and liabilities. To mention this point just here seems to be important in view of the legitimatory problems which originate within the global context and which, as will be seen below, suffer, quite opposite to the national case, from the lack of a world-wide law giver to establish the necessary legitimacy link between the economic and the political system.

My vision is here that the licence to operate a private company and to make profits should not be understood as being granted by law *unconditionally*. Instead, entrepreneurial freedom should be understood – and this is in line with the so called *concession theory* (Parker 2002:3) – as being granted under the *legal proviso* or constraint that corporations take over, within the context of their strategic operations, a *limited* (stratgey-centered) responsibility for making societal peace more stable. Referring again to a proposal of Lorenzen (1989:53) one could speak here of the "simple profit principle" ("einfaches Gewinnprinzip") as the decisive characteristic of the corporation in a *republic*, as opposed to the purely economic principle of unlimited profit maximization valid under the doctrine of *economic liberalism* where state and society are considered to be strictly *separated*. This vision looks at company law as the proper place where the formal link is established between private economic activities and the public interest and where legitimacy is transferred from the *political* to the *economic* system. I recognize that to make this vision concrete on the level of law might carry with it severe problems yet to be resolved; but it would, nevertheless, contribute heavily to strengthen the legitimatory basis of corporate governance[1].

(4) From this republican perspective corporate ethics is to be understood as a *self-regulation device* which is intended to *support the (national) law* as the primary peacemaking institution in and between democratic societies, and this in three ways:

1. by *complying with the law*. An outstanding case is here the well-known "legality principle" of the Bosch company; part of this principle is the important clause that complying to the rules of law must not be made an object of economic calculations by managers and employees;
2. by *complementing national law* in cases where legal provisions for peaceful conflict resolution are not or not yet in existence as is often the case in globalized business. Corporate codes of conduct to fight sweatshop conditions in the sportswear industry worldwide by well-known companies, like Nike or PUMA, are good examples here;

---

[1] It is in this sense that the Senate und the House of Representatives of Minnesota/ USA dealt with a proposal to intgrate the *public* interest into company law by „imposing liability on a corporation and its board for damages caused to the public interest." See www.revisor.leg.state.min.us/bin/bldbill.php?bill=S1529.0 &s...

3. by critizising (or even opposing) *existing law* in order to bring about and support initiatives for reform where necessary to link the public interest in a better way to the behaviour of corporations. Some companies in South Africa during the Apartheid regime can serve as an example here.

(5) Time does not permit a detailed elaboration of this concept of corporate ethics, its many presuppositions and consequences. I restrict myself to a few comments:

1. There is, of course, first of all the question of what it does mean to solve conflicts "peacefully". The definition of "peace" must be such that it links *individual freedom* with the *unity of society* in order to efficiently coordinate human actions (as the focal point of my analysis). Following Lorenzen again we can define handling of a conflict as "peaceful" when the actors involved try to come to a *general* and at the same time *free consensus* via reasoning and argument. This points to corporate ethics as *discourse ethics* and marks the distinction between ethics and the use of power in its manyfold forms. The important philosophical problem arising here is, of course, how the word "peace" is to be introduced.[2] I have decided not to follow (any more) Apel's *Trancendental Pragmatism* (Apel 1973) here, but to start the argument *before* any distinction is made on the *semantic* level between "real" and "ideal" speech situations (see also Habermas 2005b:347). This implies that in a society there must exist already, and this as the outcome of a long-standing, broadly accepted and deeply rooted experience to coordinate actions *successfully*, a *pragmatic* understanding or "know how" about what the words "reasonable", "argumentation" or "peaceful" mean. Thus, it is on the *pragmatic* level and from the *participator's perspective* as the first-person normative point of view (as opposed to the *semantic* level and *observer's perspective* as the third-person descriptive point of view) that we should *reconstruct* and introduce the notion of peace. This is, at least, what Friedrich Kambartel (1989) proposed about how to introduce the word "reason" (see Hanekamp 2001:58, for a critical review). Kersting (2002:279) followed Kambartel when he introduced the term "pragmatic justification" and analyzed its pragmatic and grammatical presuppositions. The *grammatical presupposition* (which is of special importance at this point) holds that for the process of justification nothing else is available to the parties besides the well-known *grammar of their rationality*, in which the patterns for their mutual understanding and unfolding of the world are integrated. This grammar is, according to Kersting, part of the culture in which we grew up: we are born into a world which is formed by a net of notions and procedures of justification. The important consequence is, of course, that

---
[2] Following the strict methodological view of Janich (2001:152) it would be, as the first step, necessary to introduce the adjective "peaceful" by *predicating* the specific ation and then move on to the noun "peace" as a *reflexive* term.

embarking on this philosophical position makes the entire concept of corporate ethics *culture-bound*. And one can easily imagine that this approach has, in turn, severe consequences for multinational companies regarding their policy for handling intercultural conflicts. I shall come back to this important point at the end of my paper.

2. *Corporate ethics does not replace the market*. The opposite is true: corporate ethics is based on the market as an institution for the coordination of economic actions. In market economies plans and actions are not coordinated via the *intentions* of individual actors (by way of argumentation) but via the *monetary consequences* of their (profit-oriented) actions evaluated through the price-system. I think there is convincing *historical* evidence (as opposed to mere *analytical model building* as is the case e.g. in welfare economics) for the superior efficiency of market coordination (as compared to centrally planned economies) to assure productivity, welfare and progress. Market societies are, therefore, *comparatively less* prone to conflicts of interests and insofar better suited for making peace more stable. This argument refers to the (empirical) *indirect link* between economic rationality and the public interest meant by Crook above, as opposed to the *direct link* of corporate actions to the public interest which is set up by corporate ethics. In any case, I strongly hold that changing the economic system is not on the agenda as long as it is not convincingly demonstrated that there is a more efficient alternative to the market. Let me add that this does *not* imply the methodological principle that markets precede politics; quite the opposite. The market is to be regarded as being embedded in a system of rules set by politics and law; it cannot create its own normative basis.

3. It follows that the enterprise should remain, at its core, what it is constructed for, namely a *private economic actor* in a decentralized competitive market economy. Economic responsibility for the survival of the corporation is with entrepreneurs and management; it cannot be delegated to any other outside institution such as the state. This implies that private corporations are called upon to make sufficient profits, as a precondition for the survival of the enterprise in (more or less) competitive markets.

4. The critical point of our concept of corporate ethics is that this justification of the profit principle is *necessary* but *not sufficient*. This is so because, on the level of the economy as a whole, the profit principle can only be justified *in general*, e.g. if one *abstracts* from the many specific side conditions under which concrete decisions have to be made on the corporate level, decisions about corporate strategy, the means for making profits and about probable side effects and resulting conflicts of interest. This general justification can, therefore, substantiate only the *presupposition* that the profit motive is right in principle ("Richtigkeitsvermutung"). On top of that, what seems necessary for a *complete justification* is that entrepreneurial *freedom* granted by law is linked to a broader concept

of corporate *responsibility*, a concept that *transcends the pure economic dimension*. This argument in favour of corporate ethics seems to me to be compelling; this at least under the *pragmatic assumption* that we strive to make peace in and between societies more stable. Management must, then, be kept responsible for all those cases where the peaceful resolution of conflicts caused by corporate action is not successfully settled by law. It is in this sense that corporate ethics forms a *direct link* between the public interest and corporate strategy. Needless to say that this concept of corporate ethics depends on *partners* who share a culture of peaceful conflict resolution. Famous conflicts, like the Nestlé case in the seventies (Steinmann and Löhr 1988), Shell's engagement in Nigeria in the eighties or the situation in the sports apparel industry today (Hartmann et al. 2003) show that this cannot be taken for granted. A lengthy *learning process* for all partners is usually necessary to reach the stage of peaceful conflict resolution. For Nike such a learning process is now well documented in an article in the Harvard Business Review (Zadek 2004).

5. My argument so far, then, is somewhat predicated on what one may refer to as the *situational embeddedness* of corporate ethics. There are many historical constraints which must be taken into account in each specific strategic situation in order to substantiate a pragmatic judgement about whether or not management has lived up to its responsibilities, both economic and ethical. *Situational analysis* is, therefore, an important part not only for economic decisions but also for ethics management; one cannot exercise either of them in the abstract. The situational analysis must cover many aspects and requires, at the same time, sound judgements about which of the (relevant) aspects must be accepted as given and which could probably be influenced by the corporation, at what costs and within what time span. Only then one can hope to get an idea about the *limits or the range (reach)* of social responsibility of management in a given situation. The important aspects of the situation include (Margolis and Walsh 2003:293): the resources of the company, its competitive advantages and disadvantages, the specific culture and norms governing the industry and the country the firm is operating in, the attitude of the critical public towards the industry or certain companies, and so forth, and all this in a dynamic perspective. As a result of the analysis it may turn out that, instead of the corporation, the *industry association,* nationwide or even worldwide (Hemphill 2004), or the regulatory apparatus of the state is the proper *locus* for effectively handling the problem at hand. Let me remind you of the well-known "prisoners dilemma" (as a characteristic of market economies) which may prevent individual companies from acting alone on a moral conflict, and this for the purely economic reason of "free riding" of competitors. But even in cases where a conflict cannot be handled on the corporate level managers are not totally relieved from their moral responsibility. Corporate ethics requires that

corporations unfold in such cases *political* initiatives for what is called today "ethical displacement", i.e. for shifting the problem on a higher political level where a proper solution may be possible.
6. The situational embeddedness of corporate ethics gives reason to touch on an important methodological problem, namely the *relationship between contextuality and universality.* Hanekamp (2004) made the proposal to reverse the traditional methodology, which is to *first* argue for the universality of norms and *then* apply these norms in concrete historical situations. Instead of starting, as the first step, with decontextualized norms and asking only then how far these norms must be re-contextualized, Hanekamp reminds us of the alternative: How far must we go on in the process of de-contextualization of norms to solve those tasks which pose themselves in a given situation. The methodology of justification of norms should proceed "bottom up" instead of "top down". This comes close to what Kersting (2002:279) has called the *pragmatic presupposition* mentioned above. Kersting insists that there is no need to justify actions, decisions or norms per se, so to speak as a philosophical exercise and as an end in itself. The need for justification, he insists, arises for any field of human action only within the context of a specific practical situation, in the light of new problems and from the point of view of individual persons. I think it is worthwile to think in more detail about Hanekamp's proposal.

(6) There are, of course, many other problems which would need to be mentioned here. Among the most difficult one for management theory counts the design of a management system which allows to efficiently and effectively bring to bear ethical considerations on all five classical managerial functions, that is: on planning and control, on organization and personal and on leadership (Haas 1998; Steinmann and Olbrich 1998; Steinmann and Scherer 2000; Leisinger 2003 for Novartis). It is only when this problem is solved properly that corporate ethics really assumes practical relevance.

3 Corporate Ethics and Globalization

(1) Having outlined my notion of corporate ethics I now turn to some developments in the course of globalization which, in my view, could be understood as manifestations of corporate ethics, or, if you like, "international" corporate ethics. New challenges for corporations here emerge on the international or global level (Haufler 2001; Scherer 2003). Quite a few companies have taken on these challenges in different forms and ways (KPMG 2005). *Individual companies* have developed corporate ethics programs on their own, as e.g. in the sports apparel industry, in response to long standing public criticism. In this industry some companies set up social and environmental standards for suppliers in developing countries to address sweatshop con-

ditions. The *federation of the apparel industry*, national and international, is now trying to come to a *collective* agreement on these topics to cope with the free rider problem, especially of no name products. There are, moreover, what one calls today "Public-Private-Partnerships" as a kind of institutional arrangement between a nation state and private corporations. Such partnerships are manyfold, especially within the area of development aid. But what I would like to mention explicitly are the well-known *US-Sentencing Guidelines* from 1991 as an outstanding example of how private companies can help to support the public interest (Steinherr et al. 1998). These guidelines of criminal law offer a substantial reduction in fines to private companies if they have made a number of specific organizational provisions to fight crime in their sphere of influence (Ethics Officer, Code of Ethics, Training, Sanctions etc.)[3]. Both, companies and the state, instead of being opposing parties in criminal court, sit here in the same boat to fight crime. And, finally, let me mention what Kofi Annan has called "Global Public Policy Networks" as a type of public-private-partnership which I shall elaborate on a bit later on.

(2) Up to now there is no empirical evidence how many of these new "forms" of management actually exist. I think it is fair to say that what we talk about here is "work in progress", not even well understood in theoretical terms. The UN estimates that about 50 to 60 GPPNs exist. This in view of some sixty to seventy thousand transnational companies controlling approximately 800.000 affiliated organizations (UN Conference on Trade and Development 2001:6; Leisinger 2004:FN 181). Nevertheless, management theory must, I think, engage in this important practical development as early as possible, if only for stimulating interdisciplinary research. It is for this reason that some members of the German Business Ethics Network cooperate now with PUMA company, and this in the third year, acting as moderators of discourses which the company has inititated with its stakeholders on a worldwide level, in order to anticipate, clarify and analyze conflicts arising out of corporate strategy with the aim to avoid or peacefully solve them (Löhr et al. 2005). As already mentioned, Nike is engaged in similar discourses; it has, in fact, taken the lead here after years of hard confrontation with NGOs (Zadek 2004).

I originally intended to talk about our experience with PUMA; but this would deliver as yet only rather anecdotal evidence. I thought it might give you a broader perspective if I summarized some information about Global Public Policy Networks, to show that what is going on here may well be understood as a manifestation of corporate ethics.

(3) Global Public Policy Networks, not to be confused with what Habermas (2005:359) calls "Global Economic Multilaterals" (mainly World Bank,

---

[3] The development of the Sentencing Guidelines up to the present is documented in www.ussc.gov.

World Trade Organization, World Monetary Fund) have gained the special attention of the UN as important institutional arrangements for global governance. Kofi Annan referred to challenges of global political governance in his 1999-speech at the World Economic Forum in Davos and in his millenium address of 2000. In both documents he raises the more and more pressing question where in the world *new loci of responsibility* do emerge and what institutions could form part of a new global political order, institutions which are able and legitimized to solve the upcoming problems and conflicts in a globalized economy. Here is his answer (Reinicke and Deng 2000:XVIII):

> The United Nations once dealt with governments. By now we know that peace and prosperity cannot be achieved without partnerships involving governments, international organizations, the business community, and civil society.

What Annan has in mind, *inter alia,* are networks of public-private partnerships which are able to better cope with the complexity of governance problems on a global scale, better than formal institutions. Let me quote again (Reinicke and Deng 2000:XVIII):

> Formal institutional arrangements may often lack the scope, speed and informational capacity to keep up with the rapidly changing global agenda. Mobilizing the skills and other resources of diverse global actors, therefore, may increasingly involve forming loose and temporary global public policy networks that cut across national, institutional and disciplinary lines. The United Nations is well situated to nurture such informal "coalitions for change" across our various areas of responsibility.

Corporations as partners of "Global Public Policy Networks" sharing responsibility with other global players for the peaceful resolution of worldwide problems and conflicts: it is my view that this vision of the UN to improve processes of global governance implies for corporations exactly that kind of responsibility which corporate ethics would require of management.

A closer look at what Global Public Policy Networks do will underline this thesis. It will reveal that corporations present themselves here not in their capacity as strictly private actors which plan their economic calculations only, and only, towards the well-being of shareholders and management – without any real concern for the public interest. Instead, they participate in processes to promote the public interest in a direct way, thereby transcending the traditional "raison d'être" of the private corporation.

(4) I restrict myself to some basic remarks about these networks; more details are available in the report "Critical Choices, The United Nations, Networks, and the Future of Global Governance" by Reinicke and Deng (2000) on which I draw heavily in what follows.

The first thing I should mention is that GPPNs are not another attempt at *top-down* organization building. They are, instead, institutional innovations which emerge *bottom-up* to solve concrete issues or problems identified, and this usually on a temporary basis. Governments, international organizations, corporations and NGOs may be partners depending on the complementarity of resources needed to tackle the specific issue at hand. There are six *functions* networks can perform, which Reinicke and Deng discuss in some detail. Let me shortly mention three of them (Reinicke and Deng 2000:27):

1. Networks contribute to *establish a global policy agenda* and offer mechanisms for developing a truly *global public discourse* in which to debate the agenda. Transparency International (TI) is an example. The problem was and still is here to crack the taboo around corruption, without alienating the very people on whom it would rely to make inroads into the problem. An important part of the strategy is to build "islands of integrity" with corporations, public authorities of the country and TI as partners in cooperative anticorruption efforts. All relevant companies commit themselves in their code of conduct to refrain from bribery and to develop appropriate organizational structures for the implementation of the code. I think this is a fairly good manifestation of corporate ethics. The "islands of integrity" are provisions to overcome the well-known prisoners' dilemma.
2. Networks facilitate processes for *negotiating and setting global standards.* Setting transnational rules and standards is becoming ever more important as political and economic liberalization and technological change create transnational social and economic spheres of activity whose governance demands a global framework. More and more national and international bureaucracies realize that negotiating and setting standards to address transnational problems differ from agenda-setting in their need to involve all the stakeholders, both because these stakeholders provide timely and complex knowledge and because their involvement provides, according to Reinicke and Deng, legitimacy to the process through inclusion of those concerned. A good example is the "World Commission on Dams" (www.dams.org) which deserves to be outlined in some detail. It was the growing complexity and politicization of large-dam construction and its social, economic, and environmental implications which made this, according to Reinicke/Deng, one of the most conflict-ridden issues in the development debate. In the late 1980s and early 1990s, a breakdown of dialogue among NGOs, dam builders, and international organizations such as the World Bank, which had financed many large dam projects worldwide, led to a stalemate. This stalemate imposed considerable costs on all stakeholders: builders saw their income from dam construction decline; NGOs had to spend considerable resources to sustain public campaigns against large dams; and the World Bank, facing fierce public pressure, could no longer support any loans in this area. Bringing representatives from all relevant groups and sectors together in an independent *trisectoral network* was imperative to break the stalemate and to start to build a consensus on standards for large-dam construction. The World Commission on Dams did just this. The report published in 2000 gives evidence on the impressive results as a fair *normative basis* for future economic activities in this field. It

was Göran Lindahl, then president of the multinational engineering firm ABB, who supported the whole project; he realized very early that a trisectoral effort could lead to greater stability and predictability in the industry's business environment. But the list of supporters of the initiative contains even more names of important companies all over the world. I think participating in this network is a clear *manifestation of corporate ethics*. Of course, it is one thing to develop a code of conduct and another to implement it. Here lies a severe deficit of the World Commission on Dams. What is necessary is that corporate ethics becomes an integral part of every-day management; only then will companies commit themselves to actually implement this normative framework on a permanent basis and not to just confining to a mere exercise of window-dressing. And what we observe, moreover, in the sports apparel industry is that NGOs are more and more calling for independent auditors to check results.

3. Rather than bringing about a concrete result, as in these two examples, a third function of networks turns out to be the result of the *cooperative process* itself, in the sense of an important by-product, namely that GPPNs may help closing the *global participatory gap*. Reinicke and Deng rightly point out that there is no global public space in which substantive discussion of transnational challenges can effectively take place and be acted upon in an open and participatory fashion. They note that economic, cultural, and social integration requires more than simply efficient *technocratic* management. All emerging problems, be it the regulation of the Internet, solutions to preserve the ozone layer, the control of international money laundering, have also a *political* dimension. It requires *inclusive* and *legitimate* political processes which – according to the UN – Global Public Policy Networks can hopefully promote, because of their rather inclusive and discursive character. A case in point is again the World Commission on Dams network which got willy-nilly engaged in processes of determining what is or is not in the broader public interest. Within this process networks raise the profile of an issue to the point where addressing it is considered to be in the global public interest. So it is the *process* itself, not the final product, by which public issues are raised and treated in rational discourses. This suggests that participating in Global Public Policy Networks allows for a *learning process* in exercising corporate ethics. But the question remains, of course, whether this process has any legitimatory potential at all, as presupposed by some of its protagonists.

(5) Unfortunately, there is no time to go into more details about Gobal Public Policy Networks[4] (see also Benner et al. 2002). I realize that what I have said so far is still rather selective and far away from a convincing picture of business reality to really support my thesis that these emerging phenomena

---

[4] There is a serious critical discussion about the negative effects on development countries (e.g. Zammit 2003).

are manifestations of corporate ethics. In order to draw such a picture we would need, in fact, more empirical research about – generally speaking – existing *global business regulation* and the role private corporations and other institutions played and still play in developing and implementing such rules (a ground breaking work is Braithwaite and Drahos 2000). Only on such an empirical basis can we hope to fully understand the role of private actors in setting global rules and to make proposals for improvement. To indicate the difficulties arising here let me pick up in the last part of my paper a few of the research questions which my discipline, in asking for help, poses to other academic fields.

4 Open Problems and Research Questions

(1) There are, first of all, quite a few unsettled *philosophical questions*. The most difficult one is the problem of what, if at all, could count as a sound argumentative *basis for (international) corporate ethics* and ethics in general. There are, as you know, quite a few philosophical schools which seem "incommensurable" from an outside point of view. I mentioned already that our concept of corporate ethics tries to apply fundamental concepts of *Methodological Constructivism*, developed in the tradition of the former "Erlangen School", thereby following the linguistic and pragmatic turn in philosophy, as opposed e.g. to Apel's *Transcendental Pragmatism*. Our concept is thus culture-bound as mentioned earlier already. And this raises the question often discussed today under the heading of "relativism": How can one come to *universal values* in view of the plurality of cultures? This question is obviously of great importance for international management: it depends e.g. on the answer to this question how managers of international companies should interact with foreign cultures. Personally, I do not consider myself qualified to delve into the abyss of this philosophical problem. Whereas Rorty (2005) strictly defends a relativistic position still today[5], Harald Wohlrapp (1998, 2000) elaborated extensively on a non-relativistic cultural position. Moreover, it was Carl Friedrich Gethmann who made a valuable contribution here at the conference of the European Business Ethics Network in 1996. In his paper "Reason and Cultures, Life-world as the Common Ground of Ethics" (Gethmann 1998) he holds, *first*, "that the philosophy of Reason overlooks that it is actually and in any case only a plausible programme with regard to certain factual structures of the life-world" (214); and he demonstrated, *secondly*, that the contextualists overlook that it can be an immanent desideratum of particular identity to rise above the limits of the particular. Under certain conditions", he holds, "Reason itself – at least as a desideratum – is a cultural fact. But this fact would perish if the culture perished." (214). Thus, according to Gethmann, "a common basis in the life-

---

[5] For a critical analysis of Rorty's position see Lueken (1998:307).

world is necessary for universal moral convictions to be founded. [...] A uniform, universal form of life is by no means the maxim which results from the project of Reason." (214). Without going into details I think that this provides a useful orientation for further research, e.g. in International Human Resources Management. One consequence would be that expatriates should be trained not to *preach* western values as universal truths a priori but to try to initiate, if necessary, on the *pragmatic* level a *learning process* with members of other cultures, in an attempt to unfold, step by step and based on the *successful outcome* of common actions, a mutual understanding of what it means to solve conflicts peacefully (Steinmann 2004).

(2) A second unsettled problem relates to the *character and legitimacy of rules* developed by corporations – be it alone, on industry level, or in global networks, together with other private and non-private institutional actors. Legal scholars speak here of "soft law" (Shelton 2000) and have lengthy discussions about this grey zone between legal norms and social norms, between the societal articulation of interests and their transformation in formal legal rights and norms.

Klaus Günther (2001:541) has raised crucial questions about this development in a paper about globalization as a problem of the theory of law. He points out, *inter alia,* that the plurality of law generating regimes which developed worldwide as a consequence of globalization threatens an important – and so far generally accepted – principle of any democratic state, namely the *unity of the legal system*. Günther raises disturbing questions: What about justice, about the principle of equal treatment of equal cases, thus far accepted, at least in principle, as a precondition in a coherent system of *national* law? What about the autonomy of nation citizens which alone can produce legitimacy within the democratic state? How is legitimate law possible if private actors authorize themselves to self-regulate disparate arenas of social life?

Acceptable answers to these questions must obviously transcend the traditional notion of legitimacy (and accountability). Traditionally, legitimacy arises within the *political system* of the *nation state* from the "institutionalization of those discursive processes of opinion- and will-formation in which the sovereignty of the people assumes a binding character." (Habermas 1996:104). But on the global level there is, as yet, no law giver, no state apparatus and no sovereign people. Thus, formal legitimacy cannot be transferred, as I indicated above for the national case, from the political to the economic system by integrating corporate ethics into company law. Instead, we have "spontaniously emerging civil society associations and movements that map, filter, amplify, bundle and transmit private problems, needs and values" (Habermas 1996:367), acting within the economic system, not the political system. Can one apply the Habermasian "theory of deliberative democracy" (1996), which was developed for the political system, directly to the economic system to qualify such processes and their output as not only factually but also formally legitimate? This is what Scherer and Palazzo (2005) seem to propose in a recent paper. But in the absence of global sovereign and proper institutions for democratic political action on a worldwide scale there is no source from which private rules can derive a binding legitimatory character. What seems to be necessary, then, is a broader notion of legitimacy which is no longer bound to the nation state. Its centerpiece would be the *public use of reason,* and this not only as a basis of the political system but also of the economic system.

(3) Closely related to this second point is a third one, resulting from the interactive and discursive character of corporate ethics. If NGOs become more and more partners of discourses with corporations, both as providers of *technical* knowledge and by participating in *political* processes intended for establishing a legitimate basis for corporate action, then NGOs must understand and practice the new role required by corporate ethics (Löhr 2004).

NGOs get in a *role-conflict.* Thus far they act as a kind of *countervailing power* campaigning *from the outside* in the public arena in order to *force* management to change its strategy in such a way that it becomes more in line with the (perceived) interests of stakeholder groups, interests which, by the way, are often more or less regarded as legitimate *per se.* Now NGOs get involved in a strategic corporate-ethics process, based on *argumentation,* from which to derive good reasons for and to commit to action programmes which regard human rights, social standards, environmental norms and so forth as an integral part of corporate strategy. Our experience with the PUMA company shows that it is obviously quite difficult for representatives of different stakeholder groups to handle this role-conflict adequately, i.e. not to fall back during discourses on a strategy of campaigning instead of scrutinizing arguments.

Needless to say that there are other questions arising out of this factual discursive practice with NGOs which relate directly to the fundamental legitimacy problem mentioned above (Bendell 2005). Who selects the respresentatives of NGOs? What groups are authorized to take part in corporate strategy discourses? Who cares for the public interest? To whom are representatives of stakeholders accountable? What about the responsibility of management towards shareholders as laid down in many modern company laws, especially in anglo-american company law? How must rules look like to coherently integrate corporate ethics processes in legitimate corporate decision processes? Where do such rules come from to be legitimate? Is it enough that they just *emerge* out of factual social processes? These are all disturbing questions pertaining to problem of legitimacy and yet to be answered.

(4) Finally I would like to mention that corporate ethics is a challenge also for the "Theory of International Relations" (Reinicke 1998, Hellmann et al. 2003, Wolf 2005). Let me remind you that traditional theory does hold here that international relations are relations between nation states which have

established already clear-cut national interests and which contract with one another to create an international order to overcome anarchy.

This theory presupposes that nation states are sovereign in a double sense. *External* sovereignty presupposes that nation states are independent of other states in formulating and implementing national interests concerning customs, tariffs and so forth. *Internal* sovereignty presupposes that the acts of giving and implementing national law should be independent of particularistic goals of societal actors and interest groups in order to find what is in the public interest in a given situation. As we know from everyday experience globalization is a threat to both the external and the internal sovereignty. When corporations, financial and non-financial, split up their value chain and distribute it to different nation states they get more and more independent from the political power and the law of their specific home country; they get even the power to influence national law, both at home and in host countries. Moreover, when corporations get involved in global public policy networks and when they help, in line with corporate ethics, to develop and implement global rules for economic action they may also undermine the sovereignty of nation states. So, it comes as no surprise that theorists of international relations are trying now to describe and to better understand these phenomena emerging in the process of economic globalization.

5 Final Remarks

(1) Thus, there are many academic fields for which the globalization process is a challenge, in general and with regard to the public role which private actors begin to take on here. What seems to be necessary is more interdisciplinary research aiming at an *empirical* theory which allows to better understand the actual economic and ethical role of the corporation as a global player; and to allow for necessary *reforms* to make peace within and between societies more stable.

(2) It was Neil Fligstein from Berkeley who outlined such an *empirical* theory in his recent book "The Architecture of Markets, An Economic Sociology of Twenty-First-Century Capitalist Societies." Let me quote at the end of my paper a central passage from this book which highlights the role of the private corporation as an *active player* in the market, as opposed to the orthodox consensus in general economic equilibrium theory which conceptualises the corporation as a *passive actor* reacting merely to market forces (23):

> A sociological approach to market institutions makes us understand that there is not a single set of social and political institutions that produces the most efficient allocation of resources. The real issue for making markets is to create political and social conditions that produce enough stability so as to allow investments. Once these institutions are created, there are a great many ways to organize firms and markets that are compatible with making profits. Since the whole society is enmeshed in market making, it is logical to argue that many possible interventions

to produce a just and equitable society are in fact compatible with profit making. Indeed, one outcome of these interventions is to strengthen the legitimacy of market institutions.

(3) In view of such a general theory of twenty-first-century capitalism it may, perhaps, turn out that corporate ethics in the rather narrow sense defined here is only part of a broader concept of *corporate citizenship* (Scherer and Palazzo 2005; Moon et al. 2005) discussed so enthusiastically today in many countries. Or it may be regarded, as Utting (2005) proposed, as part of a still broader concept which stresses the necessity to complement peace-meal and more "*collaborative*" concepts of corporate self-regulation (corporate ethics, corporate social responsibility, corporate accountability) by all kinds of governmental and non-governmental devices of "articulated regulation" oriented more towards "*confrontation*" and innovation.

What all these proposals have in common is a problem that Philipp Selznick touched on in his book "Law, Society, and Industrial Justice" as early as 1969 when he talked about the disentanglement of citizenship from private law in modern times. The consequence was, according to him, that "citizenship" is conceived today as a status and set of responsibilities that applies only to individuals but not to corporations (quoted from Parker 2002:28).

But if we begin to *reverse* this process by intentionally bringing back corporations or – more general – private companies in the political arena we must be careful not to broaden the notion of corporate responsibility too much. If we treat not only the individual but also companies as "citizens" in the full political sense of the word, then we may loose the advantages in economic welfare which industrial societies have gained over the last centuries through what we call today *systems differentiation,* with each system having its own code of rationality. I hold, that the big corporation, as one of the most important and powerful agents of the capitalist system, should remain what it is primarily constructed for, namely a private *economic* actor rather than a political actor. Of course, corporate ethics broadens this narrow economic concept of the corporation towards political responsibilites. But this should be done – and this seems to me to be important – not in an unspecified and rather unlimited way but should be restricted to the task of peacefully handling conflicts in and between societies, conflicts which are caused by corporate strategy: restricting political responsibility of corporations to peacefully handling conflicting consequences caused by corporate strategy should be seen as the *focal point* of corporate ethics (Dubbink 2004:27; Drucker 1974:341).

Even in this restricted capacity corporate ethics can, in my view, help substantially in *developing and upholding* a culture of reason and peace worldwide. In doing so, corporate ethics would contribute to re-producing *the only and central resource* which is at the basis of any democratic and free society, a resource on which the good functioning of markets depends but which the market itself cannot create. This resource is the *motivation of individuals* to transcend mere cost-benefit calculations for enhancing private welfare and to participate, instead, in ethico-political processes to foster the public interest (Habermas 2005a:II). Big and powerful multinational corporations acting solely in the name of profit and of shareholder-value will, in the long run, contribute substantially to destroy this *motivational resource* and, thereby, cut the ground from under their own feet; this process is what Habermas coined the "colonialization of the life-world" through the economic system. Thus, corporate ethics seems to me to be, in the long run, in the best interest of private corporations and should be regarded as part of good management in a globalizing world.

References

Annan K (2000) We, the Peoples: The Role of the United Nations in the 21st Century. New York
Apel K-O (1973) Transformation der Philosophie, Band II – Das Apriori der Kommunikationsgemeinschaft. Frankfurt/M.
Bendell J (2005) In whose name? The accountability of corporate social responsibility. In: Development in Practice 15 (3+4):362–374
Benner T, Reinicke WH, Witte JM (2004) Multisectoral networks in global governance: Towards a pluralistic system of accountability. Global Goverance and Public Accountability, Government and Opposition 39 (2):191–210
Benner T, Reinicke WH, Witte JM (2002) Innovating Governance: Global Public Policy Networks and Social Standards. In: Scherer AG et al. (Hrsg) Globalisierung und Sozialstandards. München und Mering, 97–115
Braithwaite J, Drahos P (2000) Global Business Regulation. Cambridge
Crook C (2005) The good company. Economist Jan:3–18
Drucker PF (1974) Management. Tasks, Responsibilities, Practices. London
Dubbink W (2004) The Fragile Structure of Free-Market Society. The Radical Implications of Corporate Social Responsibility. Business Ethics Quarterly 14:23–46
Fligstein N (2001) The Architecture of Markets. An Economic Sociology of Twenty-First-Century Capitalist Societies. Princeton and Oxford
Gethmann CF (1998) Reason and Cultures. Life-world as the Common Ground of Ethics. In: Lange H, Löhr A, Steinmann H (eds) Working Across Cultures. Ethical Perspectives for Intercultural Management. Dordrecht et.al., 213–234
Günther K (2001) Rechtspluralismus und universaler Code der Legalität: Globalisierung als rechtstheoretisches Problem. In: Wingert L, Günther K (Hrsg) Die Öffentlichkeit der Vernunft und die Vernunft der Öffentlichkeit. Festschrift für Jürgen Habermas. Frankfurt/M., 539–567
Haas RD (1998) Ethics – A global business challenge. In: Kumar BN, Steinmann H (eds) Ethics in International Management. Berlin/New York, 213–220
Habermas J (1996) Between facts and norms: Contributions to a discourse theory of law and democracy. Cambridge, Mass.
Habermas J (2005) Vorpolitische moralische Grundlagen eines freiheitlichen Staates. In: Zur Debatte. Themen der Katholischen Akademie Bayern, 35 (2005a) I–III (Nachdruck zur Debatte 1/2004 über den Gesprächsabend mit Kardinal Ratzinger am 19. Januar 2004); zugleich: ders.: Vorpolitische Grundlagen des demokratischen Rechtsstaates? In: ders.: Zwischen Naturalismus und Religion, Philosophische Aufsätze, Frankfurt/M. 2005, 106–118
Habermas J (2005b) Eine politische Verfassung für die pluralistische Weltgesellschaft? In: ders. (Hrsg) Zwischen Naturalismus und Religion. Philosophische Aufsätze. Frankfurt/M. 2005b, 324–365

Hanekamp G (2001) Kulturalistische Unternehmensethik – ein Programm. Zeitschrift für Wirtschafts- und Unternehmensethik 2:48–66

Hanekamp G (2004) Kulturalistische Unternehmensethik. Begründung und Anwendung. In: Friesen H, Berr K (Hrsg) Angewandte Ethik im Spannungsfeld von Begründung und Anwendung. Berlin u.a., 257–273

Hartmann LP, Arnold DG, Wokutch RE (eds) (2003) Rising Above Sweatshops, Innovative Approaches to Global Labor Challenges. Westport

Haufler V (2000) A Public Role for the Private Sector. Washington D.C.

Hellmann G, Wolf KD, Zürn M (Hrsg) (2003) Die neuen Internationalen Beziehungen. Forschungsstand und Perspektiven in Deutschland. Baden-Baden

Hemphill TA (2004) Monitoring Global Corporate Citizenship. Industry Self-regulation at a Crossroads. Journal of Corporate Citizenship, Summer:81–95

Janich P (2001) Logisch-pragmatische Propädeutik. Ein Grundkurs im philosophischen Reflektieren. Weilerswist

Kambartel F (1989) Vernunft: Kriterium oder Kultur? – Zur Definierbarkeit des Vernünftigen. In: ders., Philosophie der Humanen Welt. Frankfurt/M., 27–43

KPMG Global Sustainability Services (2005) KPMG International Survey of Corporate Responsibility Reporting (http://www.kpmg.com)

Kersting W (2002) Kritik der Gleichheit. Über die Grenzen der Gerechtigkeit und der Moral. Weilerswist

Leisinger KM (2003) Opportunities and Risks of the United Nations Global Compact. The Novartis Case Study. The Journal of Corporate Citizenship 11:113–131

Leisinger KM (2004) Business and Human Rights. In: UN-Global Compact/Office of the UN High Commissioner for Human Rights: Embedding Human Rights in Business Practice. UN Global Compact Office, 50–60

Löhr A (2004) The Changing Role of NGO's for Business: Instruments, Opponents, or Professional Partners? Paper presented at the EGOS-Colloquium "CSR and Business Ethics", July 1–3, in Ljubljana

Löhr A, Steinmann H, Hengstmann R, Social Standards in the Sportswear Industry – The Case of PUMA. Paper presented at the conference "Voluntary Codes of Conduct for Multinational Corporations: Promises and Challenges". New York, May 12–15 (manuscript)

Lorenzen P (1987) Lehrbuch der konstruktiven Wissenschaftstheorie. Mannheim

Lorenzen P (1989) Philosophische Fundierungsprobleme einer Wirtschafts- und Unternehmensethik. In: Steinmann H, Löhr A (Hrsg) Unternehmensethik. Stuttgart, 25–57

Lueken G-L (1998) Relativität ohne Relativismus? Amerikanischer Pragmatismus und die Überwindung irreführender Alternativen. In: Steinmann H, Scherer AG (eds) Zwischen Universalismus und Relativismus. Philosophische Grundlagenprobleme des interkulturellen Managements. Frankfurt, 291–321

Margolis JD, Walsh JP (2003) "Misery Loves Companies": Rethinking Social Initiatives by Business. Administrative Science Quarterly 48:268–305

Matten D, Crane A (2005) Corporate Citizenship: Toward an Extended Theoretical Conceptualization. Academy of Management Review 30:166–179

Moon J, Crane A, Matten D (2005) Can Corporations Be Citizens? Corporate Citizenship as a Metaphor for Business Participation in Society. Business Ethics Quarterly 15:429–453

Parker C (2002) The Open Corporation. Effective Self-regulation and Democracy. Cambridge Reinicke WH (1998) Global Public Policy, Governing without Government? Washington D.C.

Reinicke WH, Deng F (2000) Critical Choices. The United Nations, Networks, and the Future of Global Governance. Ottawa et al.

Rorty R (2005) Is Philosophy relevant to applied ethics? Speech before the Society for Business Ethics Annual Conference, Honolulu (to be published in Business Ethics Quarterly BEQ)

Ruggie JG (2002) Voluntary Initiatives and Global Economic Governance. Keynote Address at the Carl Bertelsmann Prize 2002 International Symposium Gütersloh. Germany, September 4 (manuscript)

Scherer AG (2003) Multinationale Unternehmungen und Globalisierung. Heidelberg

Scherer AG, Baumann D (2002) Corporate Citizenship bei der PUMA AG. In: Ruh H, Leisinger KM (Hrsg) Ethik im Management. Zürich, 285–298

Scherer AG, Palazzo G (2005) The Political Role of Business in Society. CSR Seen From a Habermasian Perspective. Paper presented at the 2005 Annual Meeting of the Academy of Management (CSM division) in Honolulu/Hawai

Scherer AG, Palazzo G, Baumann D (2005) Global Rules and Private Actors: Towards a New Role of the TNC in the Global Governance. Paper presented at the 2005 Annual Meeting of the Academy of Management (SIM division) in Honolulu/ Hawai

Selznick P (1969) Law, Society and Industrial Justice. New York

Shelton D (ed) Commitment and compliance. The role of non-binding norms in the international legal system, Cambridge, U.K.

Steinherr C, Steinmann H, Olbrich T (1998) Die U.S.-Sentencing Guidelines. Eine Dokumentation. In: Alwart H (Hrsg) Verantwortung und Steuerung von Unternehmen in der Marktwirtschaft. München u. Mering, 153–204

Steinmann H (2003) Unternehmensethik und Globalisierung – Das politische Element in der Multinationalen Unternehmung. In: Holtbrügge D (Hrsg) Management Multinationaler Unternehmungen. Festschrift zum 60. Geburtstag von Martin K. Welge, Heidelberg, 377–398

Steinmann H (2004) Begründungsprobleme einer Unternehmensethik, insbesondere das 'Anfangsproblem'. Die Unternehmung 58:105–122

Steinmann H, Löhr A (1988) Unternehmensethik – eine „realistische Idee". Zeitschrift für betriebswirtschaftliche Forschung 40 (1988):299–317

Steinmann H, Löhr A (1994) Grundlagen der Unternehmensethik. 2. Aufl., Stuttgart

Steinmann H, Olbrich T (1998) Ethik-Management: Integrierte Steuerung ethischer und ökonomischer Prozesse. In: Steinmann H, Wagner GR (Hrsg) Umwelt und Wirtschaftsethik, Stuttgart, 172–199

Steinmann H, Scherer AG (2000) Corporate Ethics and Management Theory. In: Koslowski P (ed) Contemporary Economic Ethics and Business Ethics. Berlin und Heidelberg

Steinmann H, Scherer AG (2002) Betriebswirtschaftslehre und Methodischer Konstruktivismus. Was leistet das kulturalistische Programm zur Grundlegung der Betriebswirtschaftslehre. In: Gutmann M et al. (Hrsg) Kultur, Handlung, Wissenschaft. Für Peter Janich. Weilerswist, 149–181

United Nations Conference on Trade and Development (2001) World Investment Report. Promoting Linkages – Internet edition, New York/Geneva

Utting P (2005) Rethinking Business Regulation. From Self-Regulation to Social Control. Technology, Business and Society Programme, Paper Number 15 of UNRISD (United Nations Research Institute for Social Development). Genf September

Wohlrapp H (1998) Die Suche nach einem transkulturellen Argumentationsbegriff. Resultate und Probleme. In: Steinmann H, Scherer AG (Hrsg) Zwischen Universalismus und Relativismus. Philosophische Grundlagen des interkulturellen Managements. Frankfurt/M., 240–290

Wohlrapp H (2000) Die kulturalistische Wende, Eine konstruktive Kritik. Dialektik 1:105–122

Wolf KD (2005) Möglichkeiten und Grenzen der Selbststeuerung als gemeinwohlverträglicher politischer Steuerungsform. Zeitschrift für Wirtschafts- und Unternehmensethik 6:51–68

Zadek S (2004) The Path to Corporate Responsibility. Harvard Business Review, Dec.:125–132

Zammit A (2003) Development at Risk. Rethinking UN-Business Partnerships, South Centre and UNRISD, Genf

Research Priorities, Profits, and Public Goods: The Case of Drug Resistant Disease[1]

Michael J. Selgelid

Overview

In this paper I argue that drug resistance is an important ethical issue in virtue of (1) the severe – and potentially catastrophic – consequences of drug resistance, (2) the fact that drug resistance largely results from the way that medicines (and research resources) are distributed, and (3) the fact that drug resistance is a clear example of a public goods problem. Recognition that freedom from infectious disease in general and drug resistant disease in particular are public goods provides additional reasons to those traditionally appealed to by bioethicists for treating health care as something special when making policy decisions about its distribution. The fact that freedom from infectious disease should be recognized as a public good, that is, provides additional (i.e. utilitarian and self-interested) reasons to standard egalitarian and human rights reasons for removing antibiotic distribution and development from free market mechanisms. While the problem of drug resistance largely results from lack of access to existing antimicrobial medications and while a primary solution to the problem of drug resistance would be the development of new vaccines and medicines, the problem of drug resistance intersects with debate over intellectual property rights in pharmaceuticals. What does all this mean for business ethics? I argue that recently proposed incentive schemes alternative to the standard patent regime reveal that conflict between corporate profit promotion, on the one hand, and corporate respect of standard humanitarian ethical duties, on the other, is not in fact inevitable. Implementation of "pull programs" such as those recently proposed by Thomas Pogge and Michael Kremer et al. would better allow industry to meet humanitarian obligations and serve the world's most important medical needs *without* (necessarily) sacrificing profits. I conclude that pharmaceutical corporations have an ethical obligation to provide the political support necessary to put such programs into place.

---

[1] Much of this material was initially presented at the 7th World Congress of the International Association of Bioethics (IAB), Sydney, Australia, November 2004, and published in an article titled 'Ethics and Drug Resistance' in Bioethics – the journal of the International Association of Bioethics, published by Blackwell.

Background

Though infectious diseases raise some of the most important ethical issues in the context of medicine, they have historically been neglected by medical ethics disciplines in comparison with topics such as abortion, euthanasia, assisted reproduction, genetics, cloning, stem cell research, and so on. The topic of infectious disease is one of the most important topics for bioethics because (1) the historical and likely future consequences of infectious diseases are almost unrivalled, (2) infectious diseases raise difficult ethical/philosophical questions of their own, and (3) the topic of infectious disease is intertwined with the topic of justice. I have elsewhere argued this case and attempted to explain why bioethics has not paid more attention to the topic of infectious disease in general (Selgelid 2005). In this paper I examine drug resistance as a specific example of a neglected ethical issue arising in the context of infectious disease and explain why the problem of drug resistance intersects with and lends importance to existing debates over intellectual property rights (in pharmaceuticals) and business ethics.

The development of antibiotics was one of the great successes of modern medicine. Penicillin was discovered by Alexander Fleming in 1928. When a method for its mass production was later developed by Howard Florey, in 1944, it was immediately hailed as a "miracle cure" and "wonder drug". Additional antibiotic and vaccine developments, in conjunction with improved sanitation and hygiene, revolutionised medicine, providing unprecedented success in the fight against infectious disease. These developments were so promising that it became common to believe that infectious disease would soon be defeated through medical progress. In 1967, for example, U.S. Surgeon General William Stewart famously declared that it was time to "close the book on infectious diseases" and shift medical attention and research resources to chronic diseases (Garrett 1994:33). Western medicine since shifted focus to things like heart disease and cancer. The medical industry is also increasingly busy with "blockbuster" treatments for allergies and depression – and lifestyle drugs for things like baldness and impotence. In the meanwhile, almost no new classes of antibiotics have been developed since the 1960s. This is largely because there is little market incentive for industry to develop antimicrobial medications.

Consequences

With the emergence and spread of drug resistant pathogens, however, existing antibiotics are losing their power to fight off infections. Though not a new phenomenon, the problem of drug resistance is increasingly being recognized as a serious, growing threat to global public health. The magnitude of the threat is revealed by the fact that the World Health Organization

(WHO) currently considers "antimicrobial resistance to be one of the top three issues in global public health" (Knobler et al. 2003:20). The WHO's "Report on Infectious Diseases 2000 – Overcoming Antimicrobial Resistance" claims that

> Drug resistance is the most telling sign that we have failed to take the threat of infectious diseases seriously. It suggests that we have mishandled our precious arsenal of disease-fighting drugs, both by overusing them in developed nations and, paradoxically, both misusing and underusing them in developing nations. In all cases, half-hearted use of powerful antibiotics now will eventually result in less effective drugs later ... [O]nce life-saving medicines are increasingly having as little effect as a sugar pill. Microbial resistance to treatment could bring the world back to a pre-antibiotic age ... The potential of drug resistance to catapult us all back into a world of premature death and chronic illness is all too real (WHO 2000).

The worry that drug resistant disease might have such severe consequences is by no means unique to the WHO. The concern that drug resistance may realistically "plunge humanity back into the conditions that existed in the pre-antibiotic age", for example, has also been raised by numerous other reputable organisations such as the U.S. Congress Office of Technology Assessment (U.S. Congress 1995).

The potential consequences of drug resistance should reveal that drug resistance – assuming that something can be done to alleviate the problem – is a matter of ethical urgency. The return to a "pre-antibiotic era" would be catastrophic. In the meanwhile the costs of treating drug resistant infections are already enormous – and estimated to be $7 billion per year in the United States alone (Smith and Coast 2003:76).

Dynamics and Distribution

Understanding of the dynamics behind drug resistance, however, reveals that worst case scenarios are not inevitable. Because resistance results from the way that drugs are distributed, the issue is a matter of distributive justice. In addition to being a product of bacterial biology, the problem of drug resistance is a social, political, and economic phenomenon. The emergence and spread of antibiotic bacteria is driven by antibiotic usage, because antibiotics themselves select for resistant strains of disease. When a person takes with antibiotics, the antibiotics kill bacteria that are susceptible, thus allowing resistant strains to flourish in the absence of microbial competitors. The emergence and spread of drug resistant (disease-causing) bacteria in human populations is therefore a function of the extent of human consumption of antibiotics. As indicated in the WHO quotation above, drug resistance ironically results from both the over-consumption of antibiotics (usually by the rich) on the one hand, and the under-consumption of antibiotics (usually by the poor, and especially in developing countries), on the other.

Over-prescription

Part of the problem is thus the widespread overuse of antibiotics. In countries such as the US and Canada, for example, it is estimated that antibiotics are overprescribed by 50% (WHO 2000). "Antibiotics are often prescribed for viral infections, for which they have no value, and for self-limited infections that would have cleared up whether or not an antibiotic had been prescribed" (U.S. Congress 1995:11). In the United States 40% of primary care physicians prescribe antibiotics when patients present with sore throats and ear aches without first seeking laboratory confirmation that this is indicated. This kind of overprescription is largely due to the time required for diagnostic confirmation (ibid.:13). It is also partly due to patient demand for immediate treatment and physicians' fear of litigation. Patient demand and physician prescription are increasingly also influenced by aggressive marketing of the pharmaceutical industry (Smith et al. 2004:12).

Another cause of drug resistance is patients' failure often fail to complete prescribed treatment regimens. While this itself promotes resistance, matters are made worse when leftover pills are saved and then used for "self-medication" later (Levy 1992). The fact that antibiotics are poorly understood by the lay public comes into play both in the case of self-medication and when patients demand inappropriate prescriptions from willing physicians.

Farming Practice

Given that drug resistance increases with antibiotic use, it is disturbing that vast amounts of antibiotics are used in agriculture and aquaculture for prophylaxis, treatment, and growth promotion. 50% (by weight) of antibiotics are used for such purposes. Of particular concern is the common practice of adding "subtherapeutic" levels of antibiotics to the food of healthy animals in order to promote their growth. In the United States, "the amount of this subtherapeutic usage is four or five times greater [by weight] than the amount used for treatment of animal diseases" (ibid.:137). Long-term, low level exposure like this creates especially favorable circumstances for the promotion of drug resistant bacteria (ibid.). While the threat to human health from such a practice is difficult to estimate and study, the worries are that resulting resistant bacteria in animals will be passed on to humans – or that resistant genes in animal bacteria will jump to bacteria which inhabit human hosts. A growing body of evidence indicates that the dangers are real (WHO 2001). While feeding animals with subtherapeutic doses of antimicrobials used for humans has been banned in Europe and Canada, regulations are lacking in the US and elsewhere. The issue, unsurprisingly, has been highly politicized.

Under-consumption

As indicated above, the failure to complete a proper course of antibiotic treatment is one of the things that promotes the emergence of drug resistant pathogens. If treatment is prematurely terminated, then disease-causing mutant bacteria, which would have been killed off if treatment had been completed, survive and become more strongly established in the absence of microbial competitors. Treatment with the very same drugs will then be less effective. This is one reason why physicians emphasize the importance of completing our treatments by taking all the pills provided. And this is why "noncompliant patients" are so often blamed for the problem of drug resistance.

The failure to complete a full course of treatment, however, is not always the fault of the patient. "Noncompliance", according to Paul Farmer, is usually a matter of *ability* rather than *agency*: "Throughout the world, those least likely to comply are those least able to comply" (Farmer 1999:255; 2003). As stated before, the poor are most likely to get sick, and least likely to afford medical care when they do. They are also least able to complete medical treatment once they start it. Poor people in developing countries often simply cannot afford to complete treatment – especially given the high drug prices set by pharmaceutical companies. In addition to drug costs, difficulty affording time off work and the cost of (often difficult) transportation to (often faraway) clinics pose barriers to the completion of antibiotic therapy. Sometimes, according to Farmer, it is simply a matter of not having enough money to rent a donkey (Farmer 2003).

Additional barriers to access are posed by infrastructural constraints of medical systems in poor countries where drug stockouts are frequent. This is well illustrated by the prison situation in the former Soviet Union. Increased crime and incarceration came with the collapse of the Soviet Union. "Almost 1 percent of the population of Russia is imprisoned, a higher percentage than any other nation in the world" (Reichman and Tanne 2002:90). Of the 300,000 prisoners released every year, 10% have active tuberculosis and more than 80% "have been infected with latent TB". Each of the latter has a 10% chance of developing active TB later in life (ibid.:53). Of prisoners (and those being released) with active TB, about a third have multi-drug-resistant TB (ibid.:90). These astounding rates of tuberculosis infection and multi-drug resistance reflect the fact that under-funded prisons are unable to maintain a steady supply of the full range of drugs needed for the long-term treatment of TB. The sporadic partial treatment that inevitably results selects directly for multi-drug resistance. Those released are unlikely to receive much better care from the wider health care system, and so the health of their families and communities is subsequently threatened, as is global public health in general. Tuberculosis is spread in the air. Both in this case and others it must be remembered that (1) infectious diseases (drug-resistant or otherwise) have no respect for national boundaries and (2) their spread is facilitated by increased international travel and trade.

While ordinary tuberculosis can be treated with a six month course of treatment costing $10, drug-resistant tuberculosis treatment takes two years and costs 100 times as much – and "[e]ven then a cure is not guaranteed." It is thus widely acknowledged that new TB drug development is needed, as there are now "300,000 new cases per year of MDR-TB worldwide." In the meanwhile it is unfortunate that, according to the WHO, there has been "a 40 year standstill in TB drug development" (WHO 2004).

Solutions

Numerous measures for curtailing the problem of drug resistance have been recommended. Increased education of health providers and the public, control of prescription practices, improvement of infection control in hospitals, and reduction of antibiotic use in farming, for example, are important parts of the solution. Given that under-consumption of drugs by the poor is a driver of resistance, improved access to medications through price reduction or social provision is also needed. Things like increased (global) surveillance and impact assessment (so the extent of the problem can be better tracked – and appreciated) and improvements in diagnostic technology are also essential.

Of paramount importance, in any case, is the development of new drugs and vaccines. New antibiotics are needed because the power of our existing supply has increasingly declined, while there has been a dearth of new drug development for decades. Vaccines are important because they prevent infection and the need for antibiotics to begin with.

The topic of drug resistance thus directly relates to the business ethics of innovation. What ethical responsibility does the pharmaceutical industry have to further invest in research and development of much-needed new antibiotics and vaccines? That the profit motive has been insufficient to motivate this kind of innovation is revealed by the current status quo. Almost no new classes of antibiotics, recall, have been developed since the 1960s, and no new TB drugs have been developed for 40 years. The failure of the medical industry to further invest in antimicrobial development is symptomatic of a more general phenomenon known as the 10/90 divide in medical research: less than 10% of medical research resources focus on 90% of the global burden of disease, and more than 90% of medical research resources focus on 10% of the global burden of disease. These numbers simply reflect the uncontroversial fact that medical industry research largely aims at promotion of profits rather than addressing the world's most important medical needs – and that the medicines that are most profitable for industry to develop do not necessarily coincide with those that are most needed from the standpoint of global public health.

Public Goods

The fact that free-market mechanisms have failed to provide solutions to the problem of drug resistance is in many ways unsurprising. The phenomenon of drug resistance reveals that freedom from infectious disease is a public good; and the ability of free markets to provide such goods is notoriously dubious. That drug resistance is a problem of public goods is illustrated by the above discussion of the dynamics behind the emergence and spread of drug resistant disease. When one individual develops a drug resistant strain of disease because of her over- or under-consumption of drugs, this more difficult or more expensive strain of disease poses threats to other individuals who may subsequently be infected. This is thus a prototypical case of negative externalities – or harms to third parties – that are a hallmark of public goods.

The usual moral justification of free markets holds that they promote both (negative) liberty and utility. It is claimed that they promote (negative) liberty because outcomes are the result of individuals' free choices about what kinds of transactions to enter into. And it is claimed that they promote utility because they are pareto efficient.[2] A pareto efficient outcome is one where it would be impossible to make any one person better off without making another worse off. It can be theoretically proven that pareto efficient outcomes would result under ideal conditions – i.e. if the parties were rational and in possession of full information, if there were no transaction costs, and so on. Because transaction costs include externalities, however, there is no good theoretical reason to expect free markets to be pareto efficient in contexts involving public goods. The status quo (e.g., as discussed two paragraphs back), in the meanwhile, seems to show that there is no good empirical reason to think that the pharmaceutical market is efficient or utility-maximizing broadly speaking.

The fact that freedom from infectious disease is a public good provides additional reasons to the egalitarian and human rights reasons traditionally appealed to by bioethicists for treating health care as something special when making policy decisions about its distribution. Norman Daniels (1995) and others are right to argue that there are egalitarian and human rights reasons to remove health care from free market mechanisms. The fact that freedom from infectious disease is a public good, however, shows that there are straightforward economic efficiency/utilitarian reasons for doing so as well. This does not mean that health care provision should be removed from market mechanisms entirely; but, as with other public goods, some kind of gov-

---

[2] Insofar as it is plausible to think that utility-maximizing outcomes might sometimes requiring making at least someone worst off, the common idea that pareto efficiency is a good proxy for utility is dubious. For the purpose of this paper, however, it is safe to leave aside this legitimate objection to the common rationale offered in favour of free markets.

ernmental intervention in the market is needed. Below I argue that governments should intervene by implementing and funding alternatives to the current regime of intellectual property right protection – and that the pharmaceutical industry should provide political support for this kind of intervention.

Intellectual Property Rights

It is at this stage that the discussion of drug resistance intersects with the issue of intellectual property rights. The problem of drug resistance is partly caused by poor people's lack of *access* to existing drugs; and the seriousness of the problem of drug resistance is exacerbated by the fact that new drugs (to fight diseases resistant to existing drugs) are not being made *available* by the pharmaceutical industry. Issues of *access* and *availability*, in the meanwhile, have been central to debates about patents in pharmaceuticals. Critics argue that patents make drugs too expensive for poor people to access them and that patents have failed to do one of the main things they are supposed to do – i.e., provide incentives for drug companies to pursue innovation of desperately needed new antimicrobial drugs and vaccines.

The extent to which developing world populations' lack of access to AIDS drugs and other essential medications is due to the high drug prices enabled by patent protection has been questioned by two recent studies of Amir Attaran (and his colleague Gillespie-White in the earlier of the two studies) (Attaran and Gillespie-White 2001:1886–1892; Attaran 2004:155–166). Most AIDS drugs, for example, are apparently not patented in most African countries; and in most African countries only very few AIDS drugs are patented. Because deadlines for patent applications have passed, this situation is not likely to change as TRIPS comes into full force. Meanwhile, according to Attaran and Gillespie-White, access to AIDS medication in African counties is low across the board; it is not the case that populations of countries where fewer AIDS medications are patented enjoy greater access to antiretroviral treatment.

The low frequency of patents on AIDS drugs in Africa furthermore holds true for essential medications more generally according to Attaran's second study:

> in sixty-five low- and middle-income countries, where four billion people live, patenting is rare for [the] 319 products on the World Health Organization's Model List of Essential Medicines. Only seventeen essential medicines are patentable, although usually not actually patented, so that overall patent incidence is low (1.4 per cent) ... I find that patents for essential medicines are uncommon in poor counties and [that patents] cannot explain why access to those medications is often lacking (Attaran 2004:155).

Holding that poverty rather than patents is the main barrier of access to essential medications, Attaran concludes that activists should shift their energies towards problems of poverty alleviation rather than objecting so much to intellectual property right protection.

Both of these studies are open to serious criticism which is beyond the scope of this paper. Even if it were true that patents do not pose serious barriers of access to existing medications, however, it would be wrong to conclude that activists concerned with health in developing countries should therefore shift attention away from intellectual property rights in pharmaceuticals (Selgelid and Sepers 2006). The *availability* problem is an additional reason to be concerned about patents. The fact that patents have failed to promote adequate innovation is revealed by the general dearth of antimicrobial and vaccine development and the phenomenon of the 10/90 divide discussed above.

Patents may spur innovation in medications likely to yield the greatest profits, but these are generally not those which are most needed from the standpoint of global public health – especially as those in need are usually poor people in developing countries. The irony here is that patents are themselves supposed to solve a public goods problem. Knowledge – in this case knowledge about how to cure disease – is a prototypical example of a public good, and patents are supposed to promote its growth by providing an incentive that would otherwise be lacking because of the free-rider problem. Without patents, the argument goes, innovating firms would lack incentive to engage in R&D because others would be able to free-ride on their discovery investment (by making copycat products). In the context of pharmaceuticals, however, even with patents the lack of sufficient incentive remains. Alternatives to the current patent regime are apparently needed if the hope is that the drugs which are most needed will both be made available via pharmaceutical industry innovation and be made accessible to the poor.

Alternative Incentive Schemes

Alternatives to the current patent regime have recently been proposed by Michael Kremer and Rachel Glennerster, and Thomas Pogge. In both cases the proposal is that wealthy governments and donor organizations should make up-front guarantees to innovating companies that they will be rewarded for developing the pharmaceuticals that are most important from the standpoint of global health. Kremer and Glennerster's "pull program" involves advance purchase commitments: a public announcement/offering is made that x doses of a needed new drug (or vaccine) meeting specified criteria will be bought from the innovating firm at price y under the stipulation that the firm will then sell additional doses at reduced price z (Kremer and Glennerster 2004). Because a market for the needed new drug (or vaccine) would thus be guaranteed in advance, an otherwise lacking incentive to fully develop it would be put into place; and so the problem of availability would be solved as long as the purchase commitment is sufficiently attractive to begin with. The requirement regarding price reduction would, furthermore, promote affordability of the new drug and thus address the problem of access.

On Pogge's related proposal (which can also be considered a "pull program") an alternative to the standard patent scheme should be implemented whereby innovating firms would be financially rewarded as a function of the extent to which their innovations lead to reduction in the global disease burden (Pogge 2005). (This new program would be an addition to, rather than a replacement of, the existing patent regime. Because firms would be given the choice to register their products under the scheme of their choice, they need not worry that the new scheme would reduce the profitability of lifestyle drugs and so on under the current patent regime, which would continue to exist in parallel to the new scheme.) As in Kremer's proposal, an otherwise lacking incentive to direct R&D to innovation of medications which are most important from the standpoint of global health would thereby be put into place. Because pharmaceutical companies would be financially rewarded in relation to the impact their drugs have on global morbidity and mortality, they would furthermore be motivated to make the relevant new drugs as affordable and widely accessible as possible – perhaps even offering them for free. Pogge's scheme, again like Kremer's, thus explicitly aims to solve both access and availability problems.

A virtue of both of these proposals is the fact that they are practically realistic to the extent that they play on the profit motive and standard market mechanisms rather than demanding that corporations become primarily motivated by altruism/good will and rather than demanding a more radical overhaul of the pharmaceutical industry and pharmaceutical economic system. The insight of both proposals lies in finding ways to align corporate profitability with the provision of products most important from the standpoint of global health.

This is a good start, but to get off the ground such proposals would require political will and a large influx of funding from wealthy governments (which would need to commit to the financial rewarding of innovating companies on either scheme). How might this political support and funding be achieved? While pharmaceutical industry (lobbying) support of such proposals might be instrumental in bringing them into fruition, the fact that industry itself would stand to gain from their implementation is one reason to hope that industry support can be won. Though Kremer and Glennerster deny that industry is likely to spend much political clout in support of proposals like theirs, Pogge is more optimistic. If pull programs really would increase the profitability of pharmaceutical companies and also restore their (now tarnished) good image as those concerned with the improvement of human health, then why shouldn't they offer political backing?

Business Ethics

Business ethics is, of course, concerned with questions about what businesses and those working in business *should* do. If Pogge is correct, then corporations should support the implementation of alternative incentive schemes like his at least partly because implementation of such schemes would open up new markets and provide additional avenues for profit making without interfering with existing ways of making profits in pharmaceuticals. If his economic arguments are technically sound, then there are straightforward self-interested reasons for taking action.

But there is more to it than that. Implementation of such schemes, assuming they are in fact realistic, would promote the greater good of humankind as well. It is often said that business executives are faced with conflicting duties: fiduciary duties to share-holders to maximize profits, on the one hand, are supposed to conflict with their more general duties as human beings to promote the good of humankind, on the other (Friedman 1970). One might argue that one should never take on roles involving duties that conflict with one's prior duties as a human being. If the role of business executive would involve duties that conflict with the prior duties of human beings in general, that is, then no human being should take on the role of business executive to begin with. And if share-holders have the duties of human beings to promote the good of humankind, then they should not require/expect the companies they own or the executives who work for such companies to do things that would compromise the good of humankind. To make a long story short, if profit maximizing endeavors preclude fulfilling our prior duties as human beings to promote the good of humankind, then neither share holders nor business executives should commit themselves to profit maximizing ventures to begin with.

A promising feature of the pull programs of Kremer/Glennerster and Pogge is that if they would actually work then the conflict (between duties to promote profits and to promote the good of humankind) so often assumed in debates in business ethics can be avoided to begin with. If these programs would in fact work, then making profits would go hand-in-hand with promoting the good of humankind. Above I argued that human beings should not take on roles that conflict with the prior general duties of human beings. Whether or not that argument succeeds, my final conclusion is that businesses and/or the individuals who comprise them have compelling moral reasons to support proposals like those of Kremer/Glennerster and Pogge. The usual excuse for corporations not directing their actions toward the promotion of the good of humankind is the idea that doing so would conflict with another (arguably primary) goal of corporations – i.e. to maximize profits. If the proposals of Kremer/Glennerster or Pogge would actually work, however, then that excuse falls away. We might not expect corporations to sacrifice their profitability, competitiveness, and ability to survive in order to serve humanitarian ends – but we should, ethically- and realistically-speaking, expect them to promote humanitarian ends insofar as this is compatible with profit-maximization. Assuming that the alternative innovation incentive schemes such as those we have been discussing actually would enable pharmaceutical corporations (and the individuals which comprise them) to both promote humanitarian ends and maximize profits, we should expect pharmaceutical corporations to provide political support for such schemes.

References

Attaran A (2004) How Do Patents and Economic Policies Affect Access to Essential Medicines, in Developing Countries? Health Affairs 23

Attaran A, Gillespie-White L (2001) Do Patents for Antiretroviral Drugs Constrain Access to AIDS Treatment in Africa? Journal for the American Medical Association (JAMA) 286

Daniels N (1985) Just Health Care. New York. Cambridge University Press

Farmer P (1999) Infections and Inequalities: The Modern Plagues. Berkeley, CA. University of California Press

Farmer P (2003) Pathologies of Power: Health, Human Rights, and the New War on the Poor. Berkeley, CA. University of California Press

Francis LP, Battin MP, Jacobson JA, Smith CB, Botkin J (2005) How Infectious Disease Got Left Out – And What this Ommission Might Have Meant for Bioethics. Bioethics 19, 4:307–322

Friedman M (1970) The Social Responsibility of Business is to Increase its Profits. The New York Times Magazine. September 13:33, 122–126

Garrett L (1994) The Coming Plague: Newly Emerging Diseases in a World Out of Balance. New York. Penguin

Knobler SL, Lemon SM, Najafi M, Burroughs T (eds) (2003) The Resistance Phenomenon in Microbes and Infectious Disease Vectors: Implications for Human Health and Strategies for Containment. Institute of Medicine, Workshop Summary. Washington DC. National Academies Press

Kremer M, Glennerster R (2004) Strong Medicine. Princeton, NJ. Princeton University Press

Levy SB (1992) The Antibiotic Paradox. New York. Plenum Press

Pogge TW (2005) Human Rights and Global Health: A Research Program. Metaphilosophy 36:182–209

Reichman LB, Tanne JH (2002) Timebomb: The Global Epidemic of Multi-Drug-Resistant Tuberculosis. New York. McGraw Hill

U.S. Congress, Office of Technology Assessment (1995) Impacts of Antibiotic-Resistant Bacteria, OTA-H-629. Washington, DC. Government Printing Office

Selgelid MJ (2005) Ethics and Infectious Disease. Bioethics 19, 3:272–289

Selgelid MJ, Sepers E (2006) Patents, Profits, and the Price of Pills: Implications for Access and Availability. In: Illingworth P, Schuklenk U, Cohen JC (eds) The Power of Pills. Pluto Press

Smith CB, Battin MP, Jacobson JA, Francis LP, Botkin JB, Asplund EP, Domek GJ, Hawkins B (2004) Are there Characteristics of Infectious Diseases that Raise Special Ethical Issues? Developing World Bioethics: 4, 1

Smith RD, Coast J (2003) Antimicrobial Drug Resistance. In: Smith R, Beaglehole R, Woodward D, Drager N (eds) Global Public Goods for Health: Health Economic and Public Health Perspectives. New York. Oxford University Press

World Health Organization (WHO) (2000) Report on Infectious Diseases 2000 – Overcoming Antimicrobial Resistance. Available at http://www.who.int/infectious-disease-report/2000/preface.htm

World Health Organization (WHO) (2001) WHO Global Strategy for Containment of Antimicrobial Resistance. Available at www.who.org

World Health Organization (WHO) (2004) Drug Resistant Tuberculosis Levels Ten Times Higher in Eastern Europe and Central Asia. Press Release, available at: http://www.who.int/mediacentre/releases/2004/prl17/en/print.html

Ethical Issues Associated with Pharmaceutical Innovation

Margaret L. Eaton

Introduction

Corporate managers responsible for the research, development, and marketing of pharmaceutical products inevitably encounter challenging ethical issues. One main reason is that, to reach the market, drugs must be studied in animals and humans, activities that have historically been plagued with ethical dilemmas about weighing the benefits and risks to subjects and the goal of advancing medical science and, in the case of human research, issues of free and informed consent. The second main reason that ethics has special relevance to the pharmaceutical industry is related to the intrinsic nature of drug products. Drugs become intensely personal when they make the difference between life and death, health and disability. The more grave the illness, the more patients care about them. Drug products are also unavoidably unsafe – no drug can be taken without some risk of harm. The diseases that have produced the need for drug therapy also make patients more vulnerable than other consumers. Fairness issues are generated by pharmaceutical products since they are often costly and therefore create access disparities. The vital importance of these products ensures that company research, development, and marketing will be closely scrutinized by governments, physicians, patients, and the press. Such scrutiny means that any lapse in corporate ethical conduct is increasingly detected and engenders a loss of public trust in the industry.

Given the propensity with which this industry encounters ethical issues, it is important to understand their genesis and to understand some of the history that can provide lessons for the future. This chapter will serve these goals by identifying ethically relevant activities of the pharmaceutical industry and presenting examples of how these activities have impacted companies. Awareness of where the ethical pitfalls lie can end a common practice of addressing these issues only in times of emergency and can allow managers to anticipate potential problems and address them proactively.

Ethical Issues Confronting Companies in the Early 21st Century

1   Conflicts of Interest with Academia

Frequently, pharmaceutical companies seek the research expertise of academics to conduct the studies that lead to product approval. Universities welcome these research contracts since they offset ever dwindling public research support, generate publications for faculty, and create training experiences for students. Despite the mutual benefits, however, conflicts of interest are common in this setting since the two entities engaged in this activity are very different one from the other. Companies do not want competitors to learn of their early drug research activity and often seek to keep this information confidential. Confidentiality is also important to protect intellectual property and trade secrets. The company sponsor also needs to design the research to achieve a commercial goal. In contrast, academics operate under conditions of openness and free discussion, and faculties are accustomed to the freedom to research any matter regardless of its short term practical application. Universities also pride themselves as sources of objective knowledge and some faculty members suspect that corporate funding influences their research focus or scientific judgment about a particular technology. Too much corporate research support, it is argued, can undermine the credibility of faculty members and the public trust in the university. In addition, academics have other campus obligations – teaching, governance – that can be compromised if a faculty member spends too much time on corporate sponsored projects. Sometimes, also, industry sponsored research can be too mundane to advance science and learning. As a consequence, even though faculty may agree to conduct corporate research, universities often refuse to abide by corporate requests to maintain confidentiality, delay publication, or alter other basic academic practices (Witt 1994; Bodenheimer 2000).

Health policy researcher David Blumenthal and colleagues at Harvard University study these conflicts of interest and commitment and have demonstrated that a certain level of industry sponsored academic research is beneficial to both parties (Blumenthal et al. 1996a; Blumenthal et al. 1996b). Beyond a certain point, however, the risks seem to outweigh the benefits. When academic researchers obtain more than two-thirds of their research funding from commercial sources, they tend to be less open, lose academic credibility, and become less productive academic citizens. Other problems include intellectual property disputes, disagreements about study design integrity and data interpretation, and conflicts over company-mandated publication delays. A certain amount of delay to submit patent applications is usually tolerated but not when delay is to preserve competitive advantages such as by restricting the dissemination of unfavorable data. On the other side of the deal, companies can be unhappy with university researchers when, for instance, they assign the work to junior faculty or students, cause delays, breach confidentiality, fail to follow the study protocol, and commit research misconduct.

An example of a research collaboration that created these kinds of conflict occurred when Sandoz collaborated with academic physicians to study the cardiac effects of one of its blood pressure drugs compared to another common drug. Some of the faculty researchers alleged that Sandoz had wielded undue influence in controlling how the research results were reported. For this reason, these faculty researchers made public their refusal to have their names included as authors on the publication (Applegate

1996). At other times, the complaint is about the magnitude of corporate research funding. Research collaborations such as those between Sandoz and the Scripps Research Institute in 1992 and Novartis and University of California at Berkeley in 1998 were so large that protestors alleged that companies were taking over and subverting the missions of premier American biomedical research and academic institutions. In the Scripps case, National Institutes of Health objections and congressional investigations forced a drastic reduction in scope (Rose 1994). Faculty objections at Berkeley delayed the Novartis deal for over six months. (Rodarmor 1998). Another well known case involved the company Immune Response which sued researchers at Harvard University and UCSF over publication of data on the company's anti-HIV drug Remune (Niiler 2000). The academic researchers found a lack of effectiveness. Corporate scientists analyzed the data differently, found a clinical benefit, and withheld data from the academic researchers in an attempt to prevent their publication of negative study results. When the academics eventually published their data, the company sued claiming damages of $7 to $10 million.

Problems like this led many universities to adopt conflict of interest and conflict of commitment policies. These policies often prevent faculty from committing too much time to corporate sponsored work, dictate the permissible financial ties to industry, require control over study design, mandate disclosures of corporate funding sources, and prevent certain controls over and delays in publishing.[1] Becoming aware of the sources of potential conflict between independent scientists and corporations and dealing with them proactively is often the best means to avoid conflict in this endeavor (Moses et al. 2002). Several resources can assist the researcher in this regard, including the reports and policies of the American Association of Medical Colleges,[2] the National Health and Medical Research Council of Australia,[3] and guidelines generated by the International Committee of Medical Journal Editors.[4] Careful attention to the avoidance and management of these conflicts serve the goal of generating mutually constructive and medically beneficial research collaborations.

2 Biased Study Design and Data Interpretation

A related ethical question relates to a concern that corporate research protocol designs are manipulated or the data interpreted to achieve a result favorable to the commercialization of a drug (Bodenheimer 2000). This is an ethical issue faced by both academic researchers and those employed by industry.

---

[1] See, for example, Stanford University Faculty Policy on Conflict of Commitment and Interest. Available from URL: http://www.stanford.edu/dept/DoR/rph/4-1.html.
[2] Available from URL: http://www.aamc.org/members/coitf/start.htm.
[3] Available from URL: http://www.nhmrc.gov.au/research/general/nhmrcavc.htm.
[4] Available from URL: http://www.icmje.org/index.html.

The dilemma arises because companies need to design trials to meet their marketing needs. For instance, if the company plans to market a drug for a severe form of a particular disease, the studies will exclude animal and human subjects with mild disease. Whether such a study design is considered biased hinges on intent which is often revealed by the manner in which the company interprets and markets the data. Using the data, for instance, to claim that the drug is useful in the disease *per se* is obviously misleading. The study may also be considered irresponsible if the company knows or strongly suspects that the drug is ineffective or toxic in mild disease and that unsuspecting physicians will tend to do what they always do – use the drug in all forms of the disease. The company may respond to this situation with a range of actions:

- claim that it is someone else's responsibility to produce study data on how the drug behaves when used "off label"[5], in this case, for patients with the mild form of the disease;
- state in the labeling that the drug is useful in severe forms of disease;
- state that no data exists on how the drug behaves in mild disease;
- disclose the suspicion that the drug is not useful or safe in the wider spectrum of diseased patients;
- act on the suspicion and conduct studies to show how the drug behaves when prescribed in the real world, in this case, for patients with the mild form of the disease.

Disagreements are increasingly common about whether drug study designs intentionally enhance efficacy outcomes or conceal side effects. Merck experienced criticism of this nature after it withdrew its anti-inflammatory drug Vioxx® in 2004, a drug that was selling $2.5 billion per year. After five years on the market, the drug was shown to enhance the risk for heart attacks and strokes prompting questions about whether the company had delayed the discovery of these risks. Leaks of internal documents revealed comments by the company's vice president for clinical research that suggested she knew that Vioxx could cause cardiovascular side effects and was contemplating designing an important clinical trial to obscure this finding (Mathews 2004). In another widely publicized case, the lead researcher responsible for ImClone's anti-cancer drug was called to testify at a congressional investigation hearing and asked to address allegations that his lucrative company ties had led him to manipulate drug study data (Urancek 2002). Whether Merck or ImClone scientists had designed studies in order to heighten efficacy or mask side effects is unknown. Yet, commentary in the scientific literature indicates a long-held uneasiness about whether company sponsored drug trials are skewed to produce positive data. Others say that this conclusion is too facile – that companies sponsor only those trials where prior data support a reasonable expectation that outcomes will be positive.

---

[5] The term "off label" refers to the use of a drug for other than approved indications.

Another questioned practice is covert reporting of the same data. For instance, researchers found that meta analysis of the side effects of the antipsychotic risperidone was skewed because what appeared to be 20 separate studies was actually only 9 studies reported multiple times, often with different lead authors (Rennie 1999). Re-publication of favorable studies artificially skews the balance of opinion in favor of a drug and some researchers suspect that drug companies do this deliberately. Another criticized study design practice deals with how companies define and code adverse drug reactions. Obviously, the number and kind of ADRs reported can be influenced by company decisions, for instance, to characterize an event as disease rather than drug related. This issue arose when regulators were attempting to determine if the antidepressant SSRIs caused an increase in suicidality. Some critics accused companies of wrongly attributing all suicidal behavior to the depression the drugs were designed to treat. Subsequently, research showed that these drugs could increase suicidal behavior and black box warnings to this effect were added to the U.S. drug labels almost 16 years after the first of these drugs was marketed (Food and Drug Administration 2004).

Another related issue involves whether companies or company ties influence data interpretation or opinions about drug products (Zuckerman 2003). Recent studies have given credence to an association if not causation in this regard. For instance, in 1998, a debate existed about the safety of calcium-channel antagonists. One study at the time showed that 96% of medical journal authors writing on the topic who supported the use of the drugs had a financial relationship with the manufacturers of the drugs compared with 37% who were critical of them (Stelfox 1998). This study and others have been cited as evidence that an association with industry lessens objectivity about the benefits and risks of drugs. Another possibility, of course, is that scientists and physicians form a favorable opinion about a product which then leads to a research affiliation with a company.

Again, the possibility that conflicts of interest can lead to biased studies and opinions has led to university and journal policies requiring disclosure of industry ties. University policies, based on the belief that the amount of money determines the amount of influence, also limit the extent to which faculty can financially benefit from commercial affiliations. These policies cannot, however, eliminate the potential for bias. Sensitivity to the problem should prompt researchers to remain vigilant – to carefully review all protocols to eliminate biasing design elements and to maintain critical and objective data interpretation.

3   Selecting Drugs to Develop

Several ethical and social questions exist about the choice of drugs to develop. The first involves whether the result is worth the effort and cost. Researchers often wish to devote their time and expertise to developing drugs that will make a major difference in the lives of sick patients. The socially beneficial aspect of their work is what drives them and keeps them in the lab. In contrast, they chafe when asked to develop a chemically related drug that offers only minimal improvement over existing therapies, so-called me-too drugs. The term me-too comes from the fact that, as soon as a large market prototype drug becomes available, several other similarly active compounds immediately follow from competing companies. This kind of drug is attractive for companies to develop since they can capitalize on the research and approval of the successful prototype, develop follow-ons more efficiently, and take an easier share of a large established market. Me-too drugs are also developed to preserve a market for a blockbuster drug, the patent for which is about to expire. Patenting a similar drug can allow the company to maintain premium pricing rather than watch revenues decline when the drug losses patent its patent and becomes generic.

Most often, the follow-ons offer some benefit over the prototype, such as lesser toxicity, improved pharmacokinetic profile, more favorable compatibility with co-morbid conditions, or easier use. Patients benefit when the me-too offers significant medical benefit and generates price competition (Dimasi 2004). The reason that many researchers dislike developing me-toos, however, is that too many offer only a marginal benefit, so small, claim some, that the intellectual and financial development resources expended far outweigh any medical or social benefit. The situation worsens, claim some, when aggressive marketing obscures the small differences between a high priced new drug and a cheaper generic. Scientists and physicians prefer intrinsic medical value rather than marketing to determine which drugs get prescribed. The fact that these drugs are considered duplicative and wasteful of resources incites consumer groups as well. Criticism of me-too drugs is relevant even in the situation where the existence of the me-too is the result of a development race rather than *post hoc* imitation. Once a strong class breakthrough drug has been approved, some favor a system that will allow only those follow-ons that demonstrate significant improvements over the prototype. Currently, most regulatory systems approve drugs based on their intrinsic merits rather than whether they offer significant benefits over existing drugs.

An example of the me-too phenomena is captopril, the prototype for angiotensin converting enzyme (ACE) inhibitors, a class of drugs used to control hypertension. The approval of captopril was followed by at least 15 me-too ACE inhibitors most of which have equivalent efficacy and side effect profiles. The similarities are so close that one pharmacologist suggested that physicians should routinely prescribe whatever brand was cheaper (Garattini 1997). An example of a drug developed to preserve revenues from a patented blockbuster is the acid reflux drug Nexium® by Astra Zeneca. This drug is the S-isomer of the company's blockbuster drug Prilosec® which had been a major revenue producer until its U.S. patent expired in 2001. Astra Zeneca took advantage of the fact that isomeric forms are considered by the FDA to be different compounds and can be separately patented. The company studies showed the S-isomer to be somewhat more effective than the racemic mix and, while other companies can experience a loss of up to 85% of a drug's revenues within a year of patent loss, Astra Zeneca was able to preserve its franchise for Prilosec by marketing Nexium, the sales for which were $3.3 billion in 2003. However, this move created a storm of criticism when patients understood that they paid up to eight times as much for the patented Nexium as for the generic and very similar (and now over-the-counter) Prilosec. Company reputation suffered as a result and a class action lawsuit was filed alleging that deceptive advertising of a medically significant benefit over Prilosec led to the demand for Nexium.[6] Whether prescribing a me-too based on price or prescribing an equally effective generic drug, the cost saving can be substantial and, at a time when financial resources for health are limited, economic considerations are not a minor point. The profusion of these follow-on drugs has led commentators to ask whether this conventional form of pharmaceutical business competition serves society as well as it should.

This me-too value issue relates to a larger question about the role of the pharmaceutical industry in society and whether the industry has an obligation to focus their efforts on innovative therapies and those for neglected illnesses. Some dislike the question since it implies a social obligation that is more appropriate for governments than business. The criticism seems aimed more at the existence of the capitalist system which would become seriously eroded if companies were required to compromise their focus on the bottom line to benefit public health. In contrast, the focus on corporate profits contributes to serious social injustice since drugs are not developed for diseases that exist in poor countries (Hartog 1993). Because of increasing pressure from underserved groups, this issue will continue to be debated by the industry and policy makers.

4 Human research ethics

Because of a long-standing history of abuse, the ethical issues associated with human research have received widespread attention. Multiple national and international guidelines and regulations were developed to ensure that human subjects are treated ethically and responsibly.[7] Because these resources are readily available, this section will deal only with the general categories of research topics that generate the most ethical concern.

---

[6] See Prescription Access Litigation Project at URL: http://www.prescriptionaccesslitigation.org.
[7] See generally, the Nuremberg Code (available at http://ohsr.od.nih.gov/guidelines/nuremberg.html), The Declaration of Helsinki (available at http://www.wma.net/e/policy/b3.htm), The Belmont Report: Ethical Principles and Guidelines for the Protection of Human Subjects of Research, Report of the National Commission for the Protection of Human Subjects of Biomedical and Behavioral Research, 44 Fed. Reg. 23,192, Department of Health, Education. & Welfare, 1979, April 18.

4.1 Risk vs Benefit

The first consideration for any researcher is to assess whether the potential benefit of the proposed research outweighs the potential harm. Harms can include physical, psychological, social, or economic. In addressing this issue, most research can be justified if benefit and risk is applied broadly in a utilitarian sense, i.e., any harm to the relatively few subjects is justified either because the research leads to new medical products that can benefit the many or, with failed trials, the benefit from medical knowledge gained and the sparing of the many patients from a useless or unsafe drug. However, ethical norms require that the risk-benefit calculus consider primarily the effects of the research on the human subjects and, additionally, on the potential to advance medical knowledge. For instance, in the U.S., IRBs (the institutional review boards that approve human research) are instructed to assess risk and benefit as follows:

> Risks to subjects [must be] reasonable in relation to anticipated benefits, if any, to subjects, and the importance of the knowledge that may reasonably be expected to result. In evaluating risks and benefits, the IRB should consider only those risks and benefits that may result from the research (as distinguished from risks and benefits of therapies subjects would receive even if not participating in the research). The IRB should not consider possible long-range effects of applying knowledge gained in the research (for example, the possible effects of the research on public policy) as among those research risks that fall within the purview of its responsibility.[8]

The Declaration of Helsinki (section A.5) is more narrow and states that the well being of the human subject should take precedence over the interests of science and society. Other aspects of minimizing risks to subjects require that sufficient prior research should be performed to understand the potential risks as much as possible. Researchers also must be qualified and adhere to standards of good research practice so as to preserve the benefits of the research as much as possible, thus not exposing subjects to risk for no good purpose. Minimizing the risks and burdens to subjects in these ways adheres to the ethical norm of non-malfeasance or doing no harm, which is a principle obligation of any research physician.

4.2 Informed Consent

Over the years, lack of informed consent has been at the heart of most unethical studies. Failure to fully inform subjects uses people as a means to an end (to produce data), disrespects the right to personal autonomy, and offends our sense of fairness. A commonly cited example occurred in a San Antonio, Texas study in the 1970s that enrolled mostly poor Mexican-American women who had had multiple pregnancies and were seeking contraceptive advice or medication. The double blind, randomized, placebo controlled study was intended to evaluate the cause of oral contraceptive side effects.

---

[8] 21 Code of Federal Regulations sec. 111(a)(2).

Subjects enrolled in the study were not told that they could receive a placebo and, as expected, there were a high number of unplanned pregnancies in the placebo group (Veatch 1971). Ethical lapses such as this are the reason for mandatory requirements that subjects be provided a clear and accurate explanation of risks, benefits, and alternatives before enrolling in any clinical study.[9] In order for consent to be meaningful, the potential subject must be given understandable information, informed of all of the expected consequences that a reasonable person would want to know about, and given an opportunity to ask questions and an adequate time to reflect before committing. Special attention to the consent process is needed when studies are complex or particularly risky and when subjects' ability to understand is compromised. Some researchers in these situations have required subjects to pass a comprehension test before enrolling them into studies.

4.3   Free Consent

Informed consent is not the same thing as free consent. Respect for persons requires that consent to research be truly voluntary. For various reasons, some patients are more vulnerable than others and are therefore more susceptible to the influence of a trusted physician who asks for consent to join a drug study. Especially when patients are desperate for any chance of treatment, physician researchers need to be careful to minimize the possibility of coercion or undue influence in the recruitment process. Vulnerable patients also include those who are economically and medically disadvantaged and who may volunteer simply because money is offered or the fact that research participation offers the only access to medical care. Those who lack mental capacity and cannot competently consent to research need a proxy who represents the patient's best interest. In addition, many believe that patients with mental deficits should be included in drug research only if the mental condition is a necessary characteristic of the research population. Most countries also have provisions forbidding or controlling the use of prisoners in research because of the difficulty in dissuading the incarcerated from the belief that cooperation with researchers will result in favorable prison treatment.

In general, little regulation exists regarding the methods or safeguards that should be employed to ensure that vulnerable or incapable subjects consent freely. As a result, individual researchers need to tailor their recruitment methods to minimize any coercive influences.

4.4   Privacy

The information collected in drug research is medical in nature and, as such, must be kept confidential. Also, the fact that a subject has entered a clinical study often reveals his or her diagnosis, so enrollment is also private infor-

---

[9]   The FDA regulations on informed consent are at 21 Code of Federal Regulations sec. 50.20 *et seq.*

mation. Every precaution should be taken to respect the privacy of the subject and the confidentiality of the patient's information.

4.5 Research in Children

Pediatric drug research has been generally neglected because of the reluctance to subject children to unknown risks given that they cannot protect their own interests through informed consent. In addition, the protective instincts of parents and guardians make them unwilling to enroll their children in research. The resultant lack of data on how children react to drugs means that most pediatric treatment proceeds through trial and error. Regulatory agencies are trying to correct this disadvantage by offering drug companies inducements to conduct pediatric research and by specifying the ethical requirements necessary to enroll these subjects in research. For instance, the FDA regulations list the safeguards that must be satisfied depending on the risk of the research, the potential benefit to the child, the potential to produce generalizable knowledge about the child's disease, and the opportunity to understand, prevent, or alleviate a serious problem affecting the health or welfare of children. The regulations also specify the qualifications of parents or other consent providers and require the assent of the child if feasible.[10]

4.6 Placebo Controls

The ethical debate about the use of placebo controls centers on the scientific benefits of eliminating the most study variables *vs* the risks to subjects of forgoing standard treatment. The use of placebo controls, according to most researchers, best satisfies the fundamental research goal of producing scientifically accurate and dependable data that can distinguish the effect of the drug from other influences, such as spontaneous change in the course of the disease, placebo effect, biased observation, or chance. Regulatory agencies will not approve a drug without sufficiently controlled data and the FDA especially adheres to the belief that no control is better than placebo.[11] Thus, drug companies often rely on the use of placebo controls because the resulting study data are considered trustworthy by the regulatory agencies and by physicians. Placebo controls are also desirable since, by producing fewer confounding variables, studies can be conducted more rapidly with smaller populations than clinical trials using active controls. Shorter clinical trials with fewer subjects can save time and money and put effective products on the market more quickly (Harrington 1999; Jost 2000; Temple 2000).

The problem with placebos is that human subjects must agree to random and blind assignment to a therapeutically inert substance masked as an active treatment. Obviously, the sicker the patient, the greater the potential

---

[10] See 21 Code of Federal Regulation sec. 50.50-50.56.
[11] In the United States, the regulations that specify the requirement for reliable study data are found at 21Code of Federal Regulations sec. 314.126.

harm from placebo administration. The duplicity (even though consented to) and the fact that diseased subjects receive no active treatment is considered unethical by some researchers and standards bodies. For instance, the World Medical Association (in the Declaration of Helsinki, C. 29) states that placebo controls should be used only in the absence of existing proven therapy. With few provisos, the Declaration requires that, if other therapies are available, experimental drugs should be tested against the best of them. The World Medical Association's stance on the use of placebos was promulgated after some controversial AIDS research in Africa and was intended to reflect a widespread interest by Association members to place subject safety over concerns about data certainty.

In the African studies, the drug AZT was tested to prevent the transmission of HIV infection from a pregnant woman to her newborn. The studies were carried out in 11 impoverished developing countries. Women in these countries were given a shorter and less expensive AZT treatment regimen than was used effectively in Western countries. The studies, which were double-blind and placebo-controlled, ultimately determined that the new treatment protocol was better than no treatment but not as effective in preventing infant infection as was the full Western treatment. According to the researchers, the use of placebos was justified because the study women had no access to prenatal care and, if they did, could not afford (nor could their countries afford) the drug as it was prescribed in the West. Thus, no study women were denied access to any locally available treatment. Plus, the studies offered at least some chance that these women would receive active treatment (Bayer 1998). Heated debates among medical professionals and medical ethicists followed publication of the study results (Angell 1997; Resnik 1998; Perinatal 1999). The major objections included the exploitative nature of the research and doubts about whether there was full informed consent (always an issue since many people do not understand the concept of placebo control even after an explanation). But the primary ethical objections focused on the fact that the HIV infection rate in the babies born of the study subjects would have been significantly less if the study had used an active treatment control with the effective Western treatment rather than placebo.

Given the differences of opinion on this topic between regulatory agencies and medical organizations, debates about placebo controls are expected to continue. In this environment, it is important for researchers to explicitly justify the use of placebos based on compelling medical and scientific criteria.

4.7 Adherence to Protocol

Research protocols need to be carefully designed in order to produce scientifically valid data and to protect the rights and interests of human subjects. Failure to adhere to the protocol undermines both. A widely publicized example of this kind of failure occurred in 1999 when an 18 year old human subject Jesse Gelsinger died in a gene therapy trial conducted at the University of Pennsylvania. Gelsinger's death resulted in investigations that disclosed multiple failures to follow protocol requirements, including failure to stop the study if subjects developed certain levels of toxicity, enrolling subjects who failed to meet eligibility criteria, failing to report side effects, and failing to notify the IRB of protocol changes. In Gelsinger's case it was likely that these failures contributed to his death and the family sued as a result. In addition, the FDA ordered Penn to halt all gene therapy experiments at its Institute for Human Gene Therapy and brought debarment proceedings against its director (FDA 2002). The tragedies associated with this incident were many and, in addition to the death, included a major setback in the progress of gene therapy research and the career injury suffered by one of the country's best gene therapy scientists. Most of the time, protocol violations do not produce such widespread destruction. But at a minimum, the failure to adhere to protocol requirements compromises the integrity of the resultant study data and thus undermines all subsequent research and clinical decisions based on that data. Consequently, it is difficult to underestimate the importance of this research requirement.

4.8 Monitoring

Vigilant study monitoring is required so that studies can be stopped as soon as the balance of harms and benefits becomes unfavorable. Monitoring has become more difficult as companies have been conducting multi-center studies often managed by independent contractors (called CROs, clinical research organizations). Hoechst Marion Roussel[12] experienced a problem in this regard when it conducted trials on Cariporide®, a drug expected to reduce cardiac tissue damage after a heart attack. The Naval hospital in Argentina was one of 26 Argentinean study sites and one of 200 medical centers worldwide that participated in the 11,500 subject trials. In 1998, Hoechst and its U.S.-based CRO, Quintiles Transnational, noticed some irregularities with the Naval Hospital trial and notified local authorities. Criminal investigations led prosecutors to allege that 137 patients had not consented to the study treatment and signatures on at least 80 consent documents had been forged. Prosecutors also collected evidence that subjects' records contained duplicated electrocardiograms with the characteristic findings needed to justify entering the patient in the trial. At least 13 subjects had died, and prosecutors claimed that some of these deaths were attributable to the experimental drug (Borger 2001). While no one thought that the drug company or the CRO had been responsible for these problems, the multitude of offshore studies and the fact that both the company and the study monitor were in other countries most likely contributed to the problem.

---

[12] This company later merged with France's Rhone-Poulenc SA to become Aventis SA.

4.9 Compensation for Injury

Most countries have requirements that research subjects be told whether they will receive medical care or compensation for injuries experienced in clinical trials. However, there is no uniform practice about the provision of treatment or compensation. Some researchers provide neither and others provide both.[13] And there is a range of practice between these two options. Injured human subjects who are not treated or compensated for their expenses and losses are left to their own devices and they sometimes sue the study sponsor. Many commentators see the unfairness in such situations and have lobbied for some uniform treatments and compensation practices. During the 1970s, a consensus began to evolve among medical ethicists that compensatory justice principles required that subjects who were injured as a result of participating in clinical research are entitled to full compensation for both medical and non-medical related costs. Policy reasoning that supported this conclusion stemmed from the view that society as a whole has a direct stake in the conduct of scientific research, including the knowledge benefits that come from experimental failure. If society shares the benefits, so should it share the financial burden of the research activity by paying for the costs of research-related injuries. Legal and ethical commentators argued that even if these injuries occurred through no fault of the researcher or sponsor, accountable public policy required that responsibility be fixed wherever it will most effectively reduce the hazards to life and health inherent in the research endeavor. If research sponsors were committed in advance to pay for injuries, the tendency to engage in excessively dangerous research would also be minimized and research sponsors would take more care in designing safe trials, monitoring for injury, and stopping trials as soon as the data revealed excessive risk. Also, companies are most often in a better financial position to bear the costs of research injury by including the cost of injury compensation in the price of the product. Finally, people would more likely consent to participate in research if they knew they would not be responsible for the costs of any injuries. Many proponents of this view also argued that since the primary goal of compensatory justice is to restore the injured person as much as possible to his or her original condition, injury compensation should include medical and non-medical costs of injury, the latter including such things as lost wages or services provided while the injured subject was disabled (Levine 1998).

These arguments have been presented over the years to various national and medical commissions (U.S. President's Commission 1982) but consistent policy has failed to result primarily because of disagreements about who should bear the expense. Consequently, researchers need to consider in advance how they will manage this aspect of their research activity.

---

[13] Companies attempt to obtain insurance to underwrite the risk but it is often not available because of the unknown risks inherent in research.

4.10 Post-study Responsibilities to Subjects

After a clinical drug study ends, subjects become patients again and are most often referred back to their primary care physician for any needed follow-up care. There are times, however, when subjects will expect something more from the company. Subjects who experience long term side effects from study drugs can expect help from the company despite any contrary provisions in the consent form. In other cases, when subjects benefit from the experimental drug, it is natural that they would want to continue taking it. Prohibitions on selling experimental drugs prevent such access and, especially when the drug was an over-all failure and will not be marketed, subjects can feel abandoned and unfairly denied effective therapy. The acuteness of the situation is heightened when the drug treated cancer or some other life threatening illness. These are reasons that the Declaration of Helsinki contains a provision that effective drugs be provided to all study subjects at the conclusion of research. Not all researchers abide by this provision and it remains a matter of debate (Lie 2004). Envisioning all possible study outcomes in advance will help researchers devise strategies to address questions about post-study support.

5 Publication of Negative Data and Data Access

Drug companies are not required to publish study results. Except under rare circumstances, regulatory agencies keep company data confidential and only require disclosure of selected summaries of study data in the drug labeling. As a result, negative drug study data often remains unpublished. Nothing stimulated an interest in changing this *status quo* as did the revelation about unpublished data on the efficacy and safety of the use of SSRIs[14] in depressed children.

All SSRI manufacturers had obtained approval to market for use in depressed adults and for years physicians had been prescribing the drugs for children off-label. Eli Lilly & Company then performed studies and obtained approval to market its SSRI Prozac® for use in depressed children. When GlaxoSmithKline applied for similar approval in the U.K. for its drug Paxil® (Seroxat® in the U.K.), the resulting investigation shed light on the phenomenon of selective publication. U.K.'s Medicines Control Agency (MCA) discovered that GlaxoSmithKline had conducted nine studies of its drug in depressed children but had published only one. When the MCA and then the FDA conducted analyses of the individual and combined data in all of the nine submitted studies, the combined data showed that the drug was not effective in pediatric depression. Furthermore, in total, these studies showed that 3.4 percent of children who were taking or had recently stopped Paxil had attempted suicide or thought more about it. That compared with 1.2

---

[14] SSRI's are selective serotonin reuptake inhibitors used to treat depression and related diseases.

percent of the children taking a placebo, a statistically significant difference. In contrast, the one study that GlaxoSmithKline had published showed that the drug was better than placebo in children and was not associated with any suicidal attempts or ideation. Further investigations revealed that other companies had also sponsored but not published negative pediatric SSRI studies. GlaxoSmithKline had not been marketing the drug for childhood depression. However, critics, regulators, and prosecutors claimed that the company knew that the drug was widely prescribed for children and that, by publishing only the positive study, the company misled physicians and patients about the efficacy and safety of the drug. The press on the story convinced the public that this kind of publication bias was common, and the ethical, clinical, and scientific integrity of the industry was called into question. The resultant furor involved the drug regulatory agencies (they started to reconsider their practice of keeping study data confidential), drug companies (they began to face lawsuits and criminal charges were brought against GlaxoSmithKline), medical journal editors (they sought wider disclosure standards), the U.S. Congress (which convened investigations), and academics and patient advocates (who demanded full access to all drug company research data) (Couzin 2004; Kennedy 2004; Letter 2004; Masters 2002; Steinbrook 2002).

The SSRI situation stimulated a renewal by national and international groups to encourage registration of all drug studies and disclosure of all drug study data (Arzberger 2004). Registries had existed for this purpose but had not been widely used by companies. Surveys had shown that existing registries had not produced the openness and transparency that would have prevented the problem seen with pediatric SSRI data. Even companies that had attempted to address publication bias had not been successful in avoiding problems. For instance, six years before the pediatric SSRI problem, GlaxoSmithKline's predecessor Glaxo Wellcome announced on its website that it would register, make accessible, and publish all of its clinical trials.[15]

Openness and access, while well intentioned, is operationally difficult. The problem with publishing all studies is that some are merely small trial balloons and small sample sizes tend to magnify or diminish influences and findings. In addition, some studies are recognized after the fact as being flawed in some important way. Publication of such studies would constitute misrepresentation of the behavior of a particular drug. Publication of nega-

---

[15] For an example of corporate trial registry, see GalxoSmithKline at http://ctr.gsk.co.uk/welcome.asp. For a collaborative industry proposal, see the Joint Position on the Disclosure of Clinical Trial Information via Clinical Trials Registries and Databases at http:// www.efpia.org/4_pos/sci_regu/Clinicaltrials2005.pdf. See also the Good Publishing Practice (GPP) for Pharmaceutical Companies at http://www.gpp-guidelines.org. The International Committee of Medical Journal Editors (ICMJE) uniform requirements for manuscripts is at http:// www.icmje.org. See also, the Consolidated Standards of Reporting Trials (CONSORT) statement at http://www.consort-statement.org.

tive studies could also lead physicians to prematurely reject a medicine. Researchers learn from failed trials and many times are able to change the study conditions to improve a drug's performance. Companies are also reluctant to publish if it undermines their ability to patent or keep sensitive business information confidential. Others wonder if the mountain of data that will become available will be digestible and practically useful for physicians and ask why physicians should not continue to rely on the regulatory agencies for the vetting of drug data. Despite these difficulties, the SSRI case stimulated renewed and wider efforts by companies, journals, and medical and scientific organizations to eliminate publication bias, allow for greater transparency and access to trial data, and improve the ability to evaluate data, flawed or not.

6 Product Pricing and Access to Drugs

New drugs are expensive because manufacturers need to recoup the considerable expense of getting a new drug to market (upwards of $800 million[16] in the U.S. at the time of this writing) (DiMasi 2001). The price paid by patients is sometimes covered by insurance or public health programs but, frequently, uninsured patients in developed countries and most patients in undeveloped countries cannot afford to pay for new drugs (Kucukarslan 1993). This disparity has prompted members of society to ask whether the industry is going to address the problem. Many large pharmaceutical companies have programs that provide needy patients with drugs and/or assist patients in obtaining public or private insurance to cover the cost of its drugs.[17] Yet, these programs help only a relatively few patients prompting countries and patient groups continue to lobbying the industry to make medicines more affordable. In the U.S., Congress frequently considers bills that would mandate drug price controls. And in Germany, where price controls exist, companies went so far as to contribute € 200 million (about $189 million in 2002) to the national health plan to prevent the country from imposing a further 4% price cut on prescription drugs (Fuhrmans 2002). Some states in the U.S. have even sued drug companies for allegedly inflating prices (Gold 2002). Patients in the U.S., where drug prices are often the highest, hire tour companies to bus them to Canada where drugs can cost as much as 50% less (Weil 2004). Activities such as this prompt debates about allowing importation of cheaper drugs, (Kennedy 2003) or controlling patent monopolies (Brown 2004).

Desperately needed drugs create extra pressure on companies as when the AIDS activist group Act-Up was created to harass Burroughs Wellcome to lower the price of the first anti-AIDS drug AZT. Act-Up's tactics were very

---

[16] Included in the cost of developing a successful drug are the costs of developing and testing drugs that never reached the market.

[17] See, for instance, Merck's Patient Assistance Program, at http://www.merck.com/pap/pap/consumer/index.jsp.

disruptive and included convincing the National Institutes of Health to adopt a "reasonable pricing rule" imposed on companies who develop drugs with the assistance of publicly funded research (Adams 2000) and prompting Congress to investigate Burroughs Wellcome drug pricing. The company did lower AZT prices and other companies with AIDS drugs were made to follow suit (Cimons 1989). Most of the pressure, however, comes from third world countries where drugs are not affordable, even for the generics used to treat common diseases such as tuberculosis and malaria. The global AIDS crisis has added fuel to the protests against drug prices and spawned multiple efforts to develop generic drugs in violation of drug company patents (Gellman 2000). All of this activity has put pressure on the pharmaceutical industry to disclose its price setting practices, lower prices, and/or relinquish patent rights to allow the manufacture of cheaper copies of patented drugs.

The pharmaceutical industry has been hard pressed to convince the public that forcing these changes, while producing short term relief, would dampen incentives for industry to engage in expensive R&D and would ultimately deprive society of valuable drugs. The U.S. Pharmaceutical Research and Manufacturers of American reported that, in 2003, its member companies invested an estimated $33.2 billion on research to develop new disease treatments. This figure amounted to spending an estimated 17.7 percent of domestic sales on R&D – a higher R&D-to-sales ratio than any other U.S. industry.[18] Without the ability to freely price, this vigorous activity will inevitably diminish. Given the high cost of development and the crucial need for prescription drugs, this cost *vs* value *vs* access debate is sure to continue for many years.

7   Marketing and Advertising

There are three main ethical issues associated with drug company marketing and advertising. The first is whether the material is truthful and fairly balances information about benefit and risk. The second issue has to do with whether companies medicalize conditions to generate a market for drugs or otherwise stimulate off label use. The third issue concerns the amount of money companies spend on advertising.

Since prescription drug regulations have been on the books, company marketing material has been required to be truthful, contain all material information, and fairly balance risk and benefit information.[18] The industry has also adopted voluntary codes to promote and support ethical marketing practices.[20] These regulations and codes are a result of continuous negotia-

---

[18] See http://www.phrma.org.
[19] The U.S. regulations for drug marketing and advertising are found at 21 Code of Federal Regulations Part 202.
[20] See, for example, Code of Pharmaceutical Marketing Practices of the International Federation of Pharmaceutical Manufacturers Associations, available from http://www.ifpma.org/.

tion between regulatory agencies, drug companies, and physician groups about the boundaries of legal and ethical prescription drug marketing and advertising practices. Nothing has tested these requirements more, however, than when companies in the U.S. were allowed to market drugs directly to patients. Until recently, it was widely believed that direct-to-consumer ads (DTC ads) for prescription drugs were inappropriate because lay people lacked the education to understand the technical medical information that was conveyed. In addition, drug ads can be misleading since they promote one drug only and fail to educate patients about other treatment options. Another primary objection concerns the fear that manufacturers, in their effort to persuade, attempt to downplay the risks and give patients the misimpression that the drugs are totally safe. The FDA struggles with DTC ads because it is under-staffed to police them and because drug companies always seem to be pushing the envelope of tolerability, especially for televised ads where the time to present risk information is limited. Physicians generally dislike DTC drug ads because they prompt patients to self diagnose and put pressure on the doctor to prescribe which often forces doctors to spend valuable time explaining why a certain advertised drug is inappropriate (Hollon 1999; Pines 1999; Terzian 1999; Wilkes 2000). For all of these reasons, most countries do not allow DTC drug advertising. In contrast, this activity has been widespread in the U.S. since the late 1980s. The drug companies claim a right to commercial free speech and the FDA, after conducting some surveys, decided that patients had a legitimate right and need to learn about newly available prescription drugs and these ads can be educational and empowering and can prompt patients to seek needed medical attention. Despite the expected benefits, however, debates continue about whether DTC prescription drug advertising is good or bad for the medical profession and the public (t'Hoen, 1998; U.S. General Accounting Office 2002).

DTC drug advertising is an effective way to increase prescribing. Research has shown that drugs that are heavily advertised, whether to physicians or patients, are prescribed more. In one study, researchers at Harvard University and the Massachusetts Institute of Technology looked at the effect of DTC drug advertising on spending for prescription drug. The study found that every $1 the industry spent on DTC advertising in 2000 yielded an additional $4.20 in drug sales. In addition, DTC advertising was responsible for 12% of the increase in prescription drugs sales, or an additional $2.6 billion, in that same year (Kaiser Family Foundation 2003). Numbers like this are the reason that the industry in the U.S. is likely to continue this activity.

The so-called medicalization issue involves the question of whether pharmaceutical company marketers intentionally (and irresponsibly) create an expansion in what is considered a disease condition. One reason, for example, that attention deficit hyperactivity disorder (ADHD) became a legitimate diagnosis was the discovery that Ritalin® could control it. A similar change occurred when drugs were found to control hair loss, male erectile dysfunction, short stature, and menopausal and pre-menstrual symptoms. This kind of expansion has been given a name by critics, diagnostic bracket creep, and has come to mean that what was once considered within the range of normal experience and behavior was now classed as illness. The introduction of new drugs is the primary reason that diagnostic bracket creep occurs. Debates exist about whether this phenomenon generates better and more enlightened treatment for patients or an irresponsible medicalization of conditions that should not be considered diseases. This debate is currently robust within the psychiatry and psychology professions when one company's SSRI antidepressant was approved and marketed for a new disease the company called social anxiety disorder (SAD). Since many people are often uncomfortable in new social situations, critics accused the company of using its ads to expand the market for its drug by blurring the distinction between a legitimate psychiatric disorder and normal personality variation (Vedantam 2001).

These criticisms of drug marketing practices have led to the question of cost *vs* benefit. If the advertising (especially the DTC kind) does not improve health awareness or provide other benefits and, worse; if the ads are harmful, the cost of advertising becomes an issue. In 2003, U.S. companies spent $3.22 billion in DTC advertising (Hensley 2004) and it is reasonable to assume that this amount is incorporated, at least to some extent, into the price of the products.[21] Supporters of DTC advertising believe that it can have a positive effect, such as when the ads prompt people to seek treatment with cholesterol lowering drugs. Advertising lipid lowering drugs, for instance, can decrease healthcare costs from the decrease in cardiovascular morbidity in treated patients. Critics, however, deplore the spending and wonder if the increase in the number of prescriptions written for advertised drugs is contributing significantly to patient health. In the midst of this debate, some companies, such as Johnson & Johnson, are skewing their patient ads to highlight the educational and safety aspects. According to the company, the move was to reduce the animosity directed to these ads and preserve the right to market directly to patients by educating and counseling consumers to improve their health (Hensley 2005).

8   Post-marketing Safety Monitoring

Drug companies have increasingly understood that the commitment they must make to a new drug is often more than the usually quoted statistic of 10 years and $800 million to get the drug to market. The post-market research and surveillance phase of a drug's life is now often as active as the pre-market phase. Companies are either obliged by the regulatory agencies or volun-

---

[21] Industry statistics indicate that the primary effect of DTC ads is to increase drug spending by increasing the number of prescriptions written, not the prices of the advertised drugs.

tarily undertake these post-market programs. Reasons for doing so are usually related to the fact that clinical research can never predict precisely how a drug will behave once prescribed in the open market. Plenty of examples exist of drugs approved for marketing only to be withdrawn within a short period because of unexpected side effects. Historically, the rate of drug recalls in the U.S. is 2–3% (Kleinke 1998). Drug companies also know that physicians will prescribe a drug for many patients who were not included on the pre-market studies – children, pregnant patients, the elderly, patients with co-morbid conditions, or even patients with other diseases. These circumstances give rise to two questions that have ethical relevance. How much post-market safety monitoring should the company perform? And, to what extent is it the company's responsibility (as opposed to the physicians) to collect data on or study off-label uses? The question about post-market safety monitoring depends on the extent of pre-market research, the utility of the drug, and the severity of potential side effects. For instance, if there was extensive pre-market research, the drug is for an unmet medical need, and the potential for side effects low, then the need for safety monitoring or further study is probably on the low side of the spectrum. The question about off-label use is more difficult to resolve. On the one hand, physicians need to take responsibility for their prescribing. If there is little data to indicate that a drug is safe for a particular use, this suggests that the physician should use the drug under research rather than clinical conditions and the company should not be responsible if the physician is reckless in this regard. On the other hand, if the company knows that the drug has a high potential for toxicity, doing nothing seems irresponsible, especially since the company benefits from the expansion in the market attributable to off-label use. This is exactly the view taken by critics in the Vioxx debate (described above). The significant legal, financial, and reputational harm that resulted in that case indicates that companies need to devote considerable attention to their responsibilities for post-market drug safety.

Obtaining Ethics Advice

Companies have several options when seeking ethics advice to manage the issues described above. Large companies tend to have in-house ethics officers who are either primarily responsible for legal compliance and report to the general counsel or, even better, are more narrowly focused on ethics and who liaison with the legal department. Resources to support such people include the Ethics Officers Association[22] and Business for Social Responsibility,[23] both of which exist to educate and enhance communication about business ethics. Some companies also empanel ethics advisory committees

---

[22] See www.eao.org.
[23] See www.brs.org.

to address ethics issues as they arise or seek outside individual consultants who work on discreet issues. Several aspects of using outsiders need to be addressed when pursuing this course of action. Ground rules and agreements must be established about confidentiality of discussions, whether the company discloses its use of ethics advice (this can seem self-serving), compensation (too much seems like buying an opinion, too little will deter participation), and how to manage ethics advice that conflicts with management decisions. The advisors themselves also need to be carefully selected. Although a company will not be well served by retaining a committed industry critic, neither is it helpful to hire a yes man who will justify and endorse all company decisions. Companies will benefit maximally from learning about all sides of a debate, and should retain an advisor who will provide objective advice while retaining a willingness to support the overall corporate effort. Expertise is also a factor. A good company advisor will be educable about the technology, knowledgeable about general pharmaceutical company business and legal affairs and about business ethics, and able to provide practical advice. Because it is rare for one person to possess all of these skills, companies may want to convene a panel of advisors with a variety of expertise (Brower 1999; Eaton 2004; Eaton 2005).

Regardless of the method used, an awareness and incorporation of an ethical perspective into business decision making will serve pharmaceutical company interests. Both economic and socials goals are advanced when business ethics programs assist in responsible development and marketing of medical products, support physicians in their treatment of patients, and advance the important social benefits of pharmaceutical and bio-technologies.

References

Adams C, Harris G (2000) When NIH helps discover drugs, should taxpayers share wealth? Wall Street Journal, B1

Angell M (1997) The ethics of clinical research in the third world. New England Journal of Medicine 337:847–849

Applegate W, Furberg K, Byungton R et al. (1996) The multicenter isradipine diuretic athersclerosis study (MIDAS). Journal of the American Medical Association 277:297–298

Arzberger P, Schroeber P, Beaulieu A et al. (2004) An international framework to promote access to data. Science 303:1777–1778

Bayer R (1998) The debate over maternal-fetal HIV transmission prevention trials in Africa, Asia, and the Caribbean: racist exploitation or exploitation of racism? American Journal of Public Health 88:567–570

Brown D (2004) Group says U.S. should claim AIDS drug patents. Washington Post, May 26, 2004, A4

Blumenthal D, Campbell E, Causino N et al. (1996) Participation of life-science faculty in research relationships with industry. New England Journal of Medicine 335:1734–1739

Blumenthal D, Causino N, Campbell E et al. (1996) Relationships between academic institutions and industry in the life sciences – an industry survey. New England Journal of Medicine 334:368–373

Bodenheimer T (2000) Uneasy alliance – clinical investigators and the pharmaceutical industry. New England Journal of Medicine 342:1538–1544

Borger J (2001) Dying for drugs: volunteers or victims? Concern grows over control of drug trials. The Guardian, February 14, 4

Brower V (1999) Biotechs embrace bioethics, BioSpace, June 14. Available from URL: http://www.biospace.com/articles/061499_ethics.cfm

Cimons M, Zonana VF (1989) Manufacturer reduces price of AZT by 20%. Los Angeles Times, September 19, 1

Couzin J (2004) Volatile chemistry: Children and antidepressants. Science 305:468–470

DiMasi JA (2001) New drug development in the United States 1963–1999. Clinical Pharmacology & Therapeutics 5:286–296

DiMasi JA, Paquette C (2004) The economics of follow-on drug research and development: trends in entry rates and the timing of development. Pharmacoeconomics 22:1–14

Eaton M (2004) Geron Corporation and the Role of Ethics Advice. In :Ethics and the Business of Bioscience. Stanford University Press, 490–511

Eaton M (2005) Affymetrix, Inc., Using Corporate Ethics Advice. In: BioIndustry Ethics. Elsevier Inc., 219–249

Food and Drug Administration FDA (2002) Notice of Opportunity for Hearing, James Wilson MD PhD, Institute for Gene Therapy, University of Pennsylvania, February 8. Available from URL: http://www.fda.gov/foi/nooh/Wilson.htm

Food and Drug Administration FDA (2004) FDA launches a multi-pronged strategy to strengthen safeguards for children treated with antidepressant medications. FDA News, October 15. Available from URL: http://www.fda.gov/bbs/topics/news/2004/NEW01124.html

Fuhrmans V, Naik G (2002) In Europe, drug makers fight against mandatory price cuts. Wall Street Journal, June 7, A1

Garattini S (1997) Are me-too drugs justified? Journal of Nephrology 10:283–294

Gellman B (2000) An unequal calculus of life and death; As millions perished in pandemic, firms debated access to drugs; Players in the debate over drug availability and pricing. Washington Post, December 27, A1

Gold R (2002) Minnesota joins list of states suing firms over drug prices. Wall Street Journal, June 19, D3

Harrington A (ed) (1999) The Placebo Effect: An Interdisciplinary Exploration. Harvard University Press; Cambridge, MA

Hartog R (1993) Essential and non-essential drugs marketed by the 20 largest European pharmaceutical companies in developing countries. Social Science & Medicine 37:897–04

Hollon MF (1999) Direct-to-consumer marketing of prescription drugs: Creating consumer demand. Journal of the American Medical Association, 281:382–384

Hensley S (2004) As drug ad spending rises, a look at four campaigns. Wall Street Journal, February 9, R9

Hensley S (2005) In switch, J&J gives straight talk on drug risks in new ads. Wall Street Journal, March 21, B1

Jost TS (2000) The globalization of health law: the case of permissibility of placebo-based research. American Journal of Law and Medicine, 26:175–186

Kaiser Family Foundation (2003) Impact of direct-to-consumer advertising on prescription drug spending, June. Available from URL: http://www.kff.org/rxdrugs/6084-index.cfm

Kennedy D (2003) When the price is wrong. Science 301:895

Kennedy D (2004) The old file drawer problem. Science 305:451

Kleinke JD, Gottlieb S (1998) Is the FDA approving drugs too fast? Probably not- but drug recalls have sparked debate. British Medical Journal 317:899

Kucukarslan S, Hakim Z, Sullivan D et al. (1993) Points to consider about prescription drug prices: an overview of federal policy and pricing studies. Clinical Therapeutics 15:726–738

Greenwood Hon. JC (2004) Letter to Bernard J. Poussot, President, Wyeth Pharmaceuticals, from Greenwood, Chairman, Subcommittee on Oversight and Investigations, U.S. House of Representatives, February 3. Available from URL: http://energycommerce.house.gov/108/Letters/02032004_1202.htm

Levine RJ (1988) Compensation for research-induced injury. In: Ethics and Regulation of Clinical Research, Yale University Press, 155–161

Lie RK, Emanuel E, Grady C et al. (2004) The standard of care debate: the Declaration of Helsinki versus the international consensus opinion. Journal of Medical Ethics 30:190–193

Masters BA (2004) New York sues Paxil maker over studies on children. Washington Post. June 3, E1[24]

Mathews AW, Martinez B (2004) Warning signs, e-mails suggest Merck knew Vioxx's dangers at early stage. Wall Street Journal. November 1, A1

Moses H III, Braunwald E, Martin JB et al. (2002) Collaborating with industry – Choices for the academic medical center. New England Journal of Medicine 347:1371–1375

Niiler E (2000) Company, academics argue over data. Nature Biotechnology 18:1235

Perinatal HIV intervention research in developing countries workshop participants. Consensus statement: science, ethics, and the future of research into maternal infant transmission of HIV-1 (1999) Lancet 353:832–835

Pines WL (1999) A history and perspective on direct-to-consumer promotion. Food and Drug Law Journal 54:489

Rennie D (1999) Fair conduct and fair reporting of clinical trials. Journal of the American Medical Association 282:1766–1788

---

[24] This article stated that the FDA had not approved Paxil for use in children, but, since physicians can prescribe drugs for non-approved uses, more than 2.1 million Paxil prescriptions for children had been written in 2002, according to the legal complaint.

Resnik DB (1998) The ethics of HIV research in developing nations. Bioethics 12:286–300
Rodarmor W (1998) Dangerous liaison? California Monthly. December. Available from URL: http://www.alumni.berkeley.edu/monthly/monthly_index/dec_98/talk.html
Rose CD (1994) Scripps deal with drug firm approved; Sandoz is partner in scaled-back alliance. San Diego Union-Tribune. May 17, A1
Stelfox HT, Chua G, O'Rourke K et al. (1998) Conflict of interest in the debate over calcium-channel antagonists. New England Journal of Medicine 338:101–106
Steinbrook R (2002) Testing medications in children. New England Journal of Medicine. 347:1462–1470
Temple R, Ellenberg S (2000) Placebo-controlled trials and active control trials in the evaluation of new treatments: part 1: ethical and scientific issues. Annals of Internal Medicine 133:455–463
Terzian TV (1999) Direct-to-consumer prescription drug advertising. American Journal of Law & Medicine 25:149
t'Hoen E (1998) Direct-to-consumer advertising: for better profits or for better health? American Journal of Health System Pharmacy 55:594–597
Uraneck K (2002) Balancing business and science at ImClone: Researchers in business struggle to manage conflicts of interest. The Scientist. December 9. Available from URL: http://www.the-scientist.com/yr2002/dec/prof_021209.html
U.S. General Accounting Office (2002) FDA oversight of direct-to-consumer advertising has limitations. Report to congressional requesters GAO-03-177, October. Available from URL: www.gao.gov/new.items/d03177.pdf
U.S. President's Commission for the Study of Ethical Problems in Medicine and Biomedical and Behavioral Research (1982) Compensating for research injuries: the ethical and legal implications of programs to redress injured subjects. Vol. 1. Department of Health and Human Services, Washington, DC. Available from URL: http://www.gwu.edu/~nsarchiv/radiation/dir/mstreet/commeet/meet16/brief16/tab_b/tab_b.html
Veatch RM (1971) Experimental pregnancy, the ethical complexities of experimentation with oral contraceptives. Hastings Center Report1:2–3
Vedantam S (2001) Drug ads hyping anxiety make some uneasy. Washington Post. July 16, A1
Weil E (2004) Grumpy old drug smugglers. New York Times. May 30, 42
Wilkes MS, Bell RA, Kravtiz RL (2000) Direct-to-consumer prescription drug advertising: trends, impact, and implications. Health Affairs 19:110–128
Witt MD, Gostin LO (1994) Conflict of interest dilemmas in biomedical research. Journal of the American Medical Association 271:547–551
Zuckerman D (2003) Hype in health reporting: "checkbook science" buys distortion of medical news. International Journal of Health Services 33:383–389

Corporate Responsibility for Innovation – A Citizenship Framework

Dirk Matten, Andy Crane, Jeremy Moon

Summary

This paper applies the metaphor of citizenship to business-society relations because citizenship is concerned with *power* and *responsibility* as two major themes in the context of ethical implications of corporate innovation. After exploring and explaining these themes, we investigate the application of the citizenship metaphor to corporations in three ways: corporations as citizens; corporations deploying government-like powers in relation to human citizens; and corporations as arenas for stakeholders to act in citizenship-like ways. We illustrate how citizenship *status*, *processes* and *entitlements* of corporations themselves, of human members of societies, and stakeholders of corporations are structured. We consider the usefulness of our approach for understanding corporate responsibility in the context of innovation as well as for future research.

1   Corporate Innovation and its Societal Impacts

The debate on the ethical implications of innovation, in particular the use of new and hazardous technologies, has a rather longstanding and well established place in the business ethics literature (di Norcia 1994). Among the papers in this volume, in particular the work of Margaret Eaton (in this volume) provides an exhaustive overview over the state of the art of the debate in particular from the perspective of business ethics. While this debate has focused in the main on applying a philosophical perspective on the issues, other contributions in this volume (e.g. the papers by Kuhlen, Nuettgens, Seiter, and Selgelid) clearly reflect a more broader debate on the wider societal and, in fact, political implications of corporate innovation and development of new technologies. Innovation and the deployment of new technologic options by private actors, most notably corporations, is increasingly perceived as having fundamental implications on the consumption choices, living standards and quality of live of individuals and societies globally. We will confine ourselves to illustrating these more recent shifts with three examples based on innovation in products, innovation in production technologies and innovation in product distribution and delivery.

- Innovation in products: one of the key challenges for pharmaceutical companies during the last years has been the question of providing access to drugs for the developing world. On the one hand, many companies have faced serious public criticism about the fact that they are unwilling to waiver their patent protection for drugs which most 'consumers' in the developing world are unable to afford. A landmark example here was the PR disaster for many major pharmaceutical companies following the (likely) defeat in a case against the South African government allowing generics for retroviral drugs to be imported into the country (Crane and Matten 2004:107–109). Pharmaceutical companies are able to provide solutions for many of the most pressing global health problems and, due to their economic power, by many are considered responsible for tackling these issues in ways which go far beyond the opportunities of governments, most notably those in the global south. On the other hand there is an increasing temptation then for pharmaceutical companies, in order to make sustained profits, to refrain from developing new products particularly catering for the diseases of the poor (Attaran 2004; Barton 2004). With considerable investment necessary to develop a new drug the risk of public outrage putting pressure on companies to then sell their products to reduced prices is just too high. Again this situation raises questions about the power of corporations to tackle fundamental global health issues.
- Innovation in production technologies: the use of genetic engineering or, more generally, biotechnological innovation has given rise to a broad debate on the general issues this raises for society at large (Harris 1994; Häyry 1994; Taunton-Rigby 2001). For instance the debate on genetically modified organisms in the food chain has put companies in the limelight and has raised issues of power and responsibility for the quality and impact of basic commodities. In particular, the ability of consumers to make an informed choice is increasingly put into question. Furthermore, as most of these innovations take place in the corporate sphere, companies are perceived increasingly as being responsible for shaping significant and often irreversible technological progress with severe implications for fundamental human rights of individuals. So, for instance, the use of the information from the genetic code of people in various products and services, from pharmaceuticals to insurance, raises questions about rules and ethical norms which should govern the development and subsequent application of these new technologies.
- Innovation in product distribution and delivery: corporations such as internet providers or mobile phone companies make use of innovative technologies to provide information and communication services. In administering these new technologies companies not only provide new products but actually shape the way fundamental consumption and life

style choices of consumers are made (Langford 2000). However, these new technologies also impact quite significantly on broader issues of societal concern. So, for instance, a debate about the corporate responsibility for child abuse based on contacts through chat rooms on the internet or by mobile phone communication has increasingly raised question about the corporate responsibility for new options of criminality facilitated by these technological innovations (BBC 2005a). More recently, the internet provider Yahoo was accused of having been directly complicit in putting a Chinese regime critic for 10 years to jail by providing crucial information about the person's email account to the Chinese prosecutors (BBC 2005b).

These are just three areas of innovation and some anecdotal examples of recent controversies around corporate innovation. Underlying these arguments is the fact that the responsibility for generating and facilitating the use of new products, technologies and services has increasingly shifted towards the private sector. Corporations thus shape increasingly fundamental consumption choices, lifestyle options and, as some of the examples show, create arenas which significantly impact on fundamental human rights of individuals. Innovation then straightforwardly raises fundamental questions about the role of corporations in society, their power and responsibility in using these technologies but also the role of governments and civil society in enabling or restricting the way business benefits from the use of innovative technology.

In the following, we will suggest a citizenship perspective as an approach to describe, analyse, understand and potentially manage some of the issues around corporate responsibility for innovation. We will do so by, first, discussing the relations between corporations and citizenship on a more general level (section 2). This discussion exposes particular shifts in business-society relations which we will explore in more detail. While section 3 then provides the basic step of applying citizenship thinking to corporations we will explore the conceptual implication of this step in section 4. The final section 5 then provides a brief conclusion and application of the suggested framework to the debate on business ethics and innovation.

2   Business–society Relations as an Arena for Citizenship[1]

This paper applies the metaphor of citizenship to business-society relations. We chose this metaphor because it raises important questions of power and responsibility which are in turn central to the developing agendas of business-society relations in the context of innovation.

---

[1] This chapter draws on our *Corporations and Citizenship* Cambridge University Press (forthcoming). It incorporates and develops our thinking in: Crane, Matten and Moon 2004; Matten and Crane 2005; Moon 1995; Moon, Crane and Matten 2005. Dirk Matten would like to thank Shahanara Bhunyan for support with the literature research for this paper.

Whilst many studies which apply the idea of citizenship to corporations to business – society relations adopt a single perspective, that of corporate citizenship or membership of society, we adopt a three dimensional perspective by analysing corporations: as if they were people-type citizens; as if they were governments in relation to people as citizens; and as if they create an arena for people to enact citizenship. Through these conceptualisations we examine the different ways in which corporations possess and structure citizenship status, entitlements and processes. We do so in the context of two important and seemingly contradictory contemporary developments in business-society relations: the nature and appropriateness of increasing business power and the new claims being made by firms about their being socially responsible.

We use the concepts of citizen and citizenship in their metaphorical sense.[2] As mentioned, we apply them to corporations in three ways: corporations as citizens who participate in political communities and authorise governments to rule therein; corporations ruling political communities through deploying government-like powers and responsibilities; and corporations creating opportunities or arenas for their stakeholders to act in citizenship-like ways. There will always be debate about citizenship's meaning, merits, and appropriateness. There are internal dynamics to this debate as new models of citizenship are developed against which practices are judged. As Marshall observed of the political concept of citizenship:

> Societies in which citizenship is a developing institution create an image of an ideal citizenship against which achievement can be measured and towards which aspiration can be measured (1950:29)

For corporations, the nature of these debates reflects social and business contexts within firms, among firms, within countries and among countries. Recognizing that, like its related political concepts, the metaphor of citizenship for corporations is essentially contested (Gallie 1956) does not, of course, obviate the need for closer investigation into its theoretical appropriateness.[3]

Corporations are generally regarded as the most prominent organisations of contemporary capitalism in part because of the employment, production, investment and wealth that they account for. They are now generally understood to be non-governmental profit-making business enterprises owned by shareholders who control the overall firm policy but managed by the agents of the owners. Their legal identity is distinct from that of their members and their internal governance regimes reflect government regulation and wider features of their national business systems. (Albert 1991; Whitley 1999)

---

[2] See Moon, Crane and Matten 2004 (432–434) for a discussion of the use of metaphors in the analysis of business.

[3] See Moon, Crane and Matten 2004 (433–434) for a discussion of essentially contested concepts.

However, numerous big businesses are known as privately-owned in that the shares in the company are not traded through stock exchanges. These remain a particularly important form of big business in parts of the world particularly Asia. Other big businesses are exclusively institutionally-owned (e.g. by banks, governments), a common form of business organisation in Rhenish capitalism of Germany, Austria and Switzerland, for example. The wider definition also brings in cooperative businesses, particularly prominent in Southern Europe. Colloquially, the word corporation is generally used to denote any form of big, private business devoted to profit-making. This is the definition that we will use as the key issues of power and responsibility link to the size, ownership and purpose of the firm, rather than to one particular feature, albeit a very important one, of ownership and control.

2.1 Business-society Relations

In the last decade or so there have been some radical developments in the agenda of business-society relations. These reflect changes in the corporations themselves and changes in the social and political context of business. Consideration of these developments has not simply been in the forums of university seminars and academic journals, nor of the pages of the financial media, nor yet in the agitprop media of the critics of corporations. The social status and impact of corporations has also been the subject of films (e.g. "The Corporation", "Roger and Me"), of documentaries (e.g. "The End of Politics", "Supersize Me") and 'airport literature' (e.g. "No Logo").

These and other forms of media coverage have brought to mass attention a whole range of issues which reflect or address business activities. For example, the role of Shell in Nigeria and the extent of its responsibilities for the social, political and economic status of the Ogoni people has raised questions about the extent to which a corporate presence is an implicit endorsement of governmental actions and the extent to which corporations should bring pressure to bear on governments. Conversely, the role of oil companies in benefiting from the US invasion of Iraq and the subsequent political settlement has animated anxieties about such close involvements with government. Recent concerns about obesity in western countries have raised the question of the role and responsibility of fast-food businesses for the health and well-being of their consumers, echoing debates about tobacco companies', governments' and personal responsibilities for cigarette consumption and attendant health risks. The publication of the 2005 Nike Social Report in which its suppliers are named and their working pay and conditions independently audited and reported represented a new landmark in the extent to which a western-based retailer is prepared to take responsibility for its supply chain.

Bringing some of these themes together we can see two simultaneous and seemingly contradictory trends. On the one hand there is a critique of what is deemed excessive business *power* such that the rights of citizens and the powers of governments (assumed in this critique to be protective of citizens' interests) are weakened. On the other hand there are claims by businesses that they are taking more *responsibility* for society, or as acting as 'corporate citizens', and there are more clearly articulated expectations by citizens and governments that corporations should take greater responsibility for society.

The view that corporations are assuming excessive power is manifest in various ways. At the level of political practice, this is evident in the anti-globalisation movement. This is a very heterogeneous movement, in terms of philosophy, organisation and tactics, but united in a main target of corporate operations across a range of countries. Their central critique is not simply that corporations have power but that this is magnified by the 'global' nature of multi-national corporations, or MNCs. In some cases anti-globalisation reflects hostility to the cultural referents of particular brands, as in the attack on McDonald's outlets in France. In other cases it reflects a critique of the business practices of major corporations through, for example, the terms and conditions of employment in the third world subsidiaries or supply chains of western clothing and sports equipment companies, the concerns of various fair trade movements. As discussed earlier, a number of aspects of the innovation process in companies and their use of technology to address, or not address, global problems is a frequent topic of critics from this movement. In other cases it reflects a general critique of the political power that goes with global economic power and the way in which this compromises the position of governments, particularly in developing countries, in their deciding the terms of inward investment of such MNCs. Hence MNCs are accused of escaping tax law, of extracting excessive benefits from developing countries and of making improper payments to secure investment opportunities. These perspectives have also been witnessed in a new literature which is critical of the activities of particular corporations and corporate activity. (e.g. Hertz 2000; Korten 1995; Monbiot 2000)

Contemporaneously, corporations have been claiming that they are acting more responsibly. Even sceptics of the idea that businesses should compromise their core market activity have noted this trend. Martin Wolf, Chief Economics Commentator of the "Financial Times", commented that there is a sense that corporate social responsibility is 'an idea whose time has come' (2002:62) and Clive Crook, Deputy Editor of "The Economist" observed that 'over the past ten years or so, corporate social responsibility has blossomed as an idea, if not as a coherent practice' (2005:3).

At the nominal level, corporations claim to be acting more responsibly through the adoption of such terms and self-descriptions as corporate citizenship, corporate social responsibly, business ethics and sustainable business. In many cases, corporations go beyond the mere adoption of sociable labels, they also seek to integrate their responsibility into their brands (e.g. BP claims to be a green energy company). Reflecting the fundamental openness to accept responsibility for the societal role of his company Jean-Pierre

Garnier, CEO of the pharmaceutical multinational GlaxoSmithKline argued in a recent interview: "I don't want to be the CEO of a company that caters only to rich countries. I'm not interested." ("Financial Times":22.VII.2005)

Whilst there may well be certain business advantages to the use of images of corporate responsibility in marketing and branding, this can also be a source of cynicism. Critics may well ask what lies behind the brand? In many cases this can be substantiated by organisational manifestations of new forms of responsibility. Many companies have now developed organisational resources and processes to reflect their increased social commitments, be it defined as corporate citizenship, corporate responsibility or sustainability. Sometimes these are free standing and in other cases they are housed in larger functional units. Some companies are assigning board level responsibilities for these new social relations. Another manifestation of new social relations is that many companies are developing programmes and policies to substantiate their commitments and organisational innovations. These range from community involvement, through concern with responsibility in the products and processes, to attention to their labour relations. Community involvement to some extent reflects a traditional commitment to philanthropy on the part of companies, whether reflecting religious and ethical commitments or more functional concerns with labour force loyalty and productivity.

However, today corporations are viewing community involvement in much more systematic rather than discretionary fashion and doing so in a way that reflects a more self-conscious stakeholder approach. Concern with the products and processes reflects a decision to ensure that goods and services reflect various social expectations in their composition, in the ways in which they are produced, and in the social and environmental externalities thereby created. Sometimes this includes securing third party audits and verification. Thirdly, many companies are also investing greater resources in workplace conditions and even in the extra-work circumstances of their staff, reflecting new demands in the area of work-life balance and new attitudes to and expectations of employment.

In some cases these new areas of company activity have been complemented by self-regulation. An obvious means to this end is through the use of external or internal corporate codes to guide and benchmark responsible behaviour of corporations and their employees. Although these are often criticised for their lack of wider accountability, they do bring opportunities for corporations to develop policies which reflect and complement their own range of commercial activities. Moreover some companies are developing codes which provide for independent verification and certification, often in collaboration with stakeholder organisations.

Another manifestation of greater company concern with their social relations is their preparedness to join business associations whose purpose is to encourage and develop the social face of business. For example, in the UK over 700, mainly large, companies are members of Business in the Community (BITC). BITC provides a variety of services and awards in the area of socially responsible business through its national and regional offices, though it emphasises that membership itself should be a step to a more reflective and proactive style of engagement with society. There are similar association in the USA (e.g. Business and Society). Internationally there are other business associations to encourage more responsible business. CSR Europe, the International Business Leaders' Forum. Membership of the UN Global Compact entails commitment to ten principles covering human right, labour standards, the environment and corruption.

Another important development has been the growth of social reporting, be it within general company communications, in dedicated social responsibility reports, or within their annual reports. Some go so far as to legitimise their reports through external verification and stakeholder engagement (e.g. British American Tobacco, Nike). Various indicators of business responsibility have also been developed and adopted in tandem. Some of these reflect agreement among corporations about appropriate reporting norms (e.g. Global Reporting Initiative).

A new burgeoning of CSR consultants suggests that companies are prepared to pay for advice about their CSR (Fernandez Young, Moon and Young 2003). There have also emerged new responsible business professional networks (e.g. CSR Chicks, Lifeworth, Association of Sustainability Professionals). A new business media on socially responsible business is also emerging. This includes dedicated media outlets (e.g. "Ethical Corporation", "Ethical Performance") as well as greater attention to these themes in the mainstream media, illustrated by the "Financial Times'" employment of a CSR correspondent and recent special supplements of corporate social responsibility in "The Economist" (22.I.2005), "The Independent" (23.III.2005) or the "Financial Times Deutschland" (7.XII.2005).

So, in conclusion, there is plenty of evidence that corporations are at least keen to be regarded as behaving more responsibly and there is also plenty of evidence of resources being invested in organisational developments consistent with this. What explains these developments?

2.2 Recent shifts in business–society relations towards the political

Having sketched something of the changing nature of business – society relations we now turn to providing some explanation for the trends that we have identified. This section is divided into two parts, the first addressing the drivers of business power and the second addressing the movement for corporate social responsibility.

Corporations are acquiring an increasingly conspicuous and, in some respects, contentious profile. There are various reasons for this. Corporations have acquired a greater share of economic participation following widespread privatisations; they have created new consumer markets; their cross border activities appear to have increased; and they have assumed greater roles in the delivery of public goods.

First, corporations have acquired more commercial opportunities. In many parts of the world this results from the waves of privatisations in what were already capitalist economies witnessed over the last quarter century (e.g. in Australia, New Zealand, the UK, the USA). Elsewhere this has resulted from more abrupt shifts following the collapse of communist regimes. As a result corporations have become responsible for more facets of citizens' lives than they used to be. In many communities, what was once delivered, for better or for worse, by governmental organisations (e.g. telecommunications, energy, water, mass transport) is now delivered by private corporations. Although governments have tended to retain regulatory, fiscal and organisational capacities, the tides of privatisation have not only had the effect of increasing the corporate sector's share of gross national product and employment but also of yielding to corporations pivotal roles in policy areas previously regarded as fundamentally political (e.g. investment in and performance of transport and utility companies; access to and use of such natural resources as water, oil and gas).

Secondly, corporations appear more conspicuous because they have created new consumer markets. This is most obviously true where there have been recent increases in the range and availability of consumer goods (e.g. China). However, it also reflects longer-term shifts in western societies from 'the politics of production' to 'the politics of consumption'. The increasing commodification of life is evident in such domestic activities as home improvements, gardening and sports.

Thirdly, corporate cross-border activities have grown. Thus, corporations are often more conspicuous simply because they are large and foreign rather than small or medium and local. This is manifest in vast increases in national foreign direct investment and international intra-and inter-company trade. This is in turn predicated on trade liberalisation facilitated by political reforms, increased access to developing economies, technological change, economies of scale and scope, and cultural homogenisation. For corporations, globalisation thereby offers opportunities to increase growth, stabilise performance, exploit new investment opportunities and increase market power.

Fourthly, there is evidence of wider changes in patterns of societal governance such that governments have reduced some modes of exercising their authority (see Moon 2002). In addition to the effect of the privatisation of governmental responsibilities in creating new market opportunities for business (see above), another corollary is that governments have actually encouraged corporations to contribute to wider governance activities. Similarly, many western companies operating in developing countries undertake such responsibilities in lieu of governmental provision be it in the provision of pensions, education, worker rights and opportunities and environmental responsibility. This expansion of corporate profile thus in part reflects regulatory failure and regulatory vacuums.

At the same time as these powerful drivers of increased business power have gathered pace other contemporary phenomena have encouraged corporations to behave more responsibly. We divide these drivers into market, social regulation and government regulation: they amount to trends towards a socialisation of markets.

A number of market drivers for more responsible business behaviour have emerged. There are new consumer demands for products and processes which reflect more socially responsible practices. Although some of the public opinion data on consumers' preparedness to punish irresponsible retailers may disguise the effect of price in their actual spending choices, certain new niche markets reflect new social values (e.g. as met by The Body Shop, Green Mountain, ethical trade systems) and of periodic occasions when consumers can be mobilised in consumer boycotts (e.g. boycotts of American sports wear companies' Bangladesh suppliers employing child labour).

There is also evidence of a greater impact of investors on the agenda of corporate social responsibility. This in part reflects the development of systems of socially responsible investment (SRI) and also the expansion of SRI agendas into wider investment criteria. Although, SRI funds still only account for a relatively small share of total investments (about 15% in the USA, 5% in the UK), these are growing and becoming more engaged with companies. Moreover, general investment funds have also taken an interest in SRI criteria, from risk and corporate governance perspectives.

Employees' expectations are also informing corporate social responsibility. This in part reflects new assumptions about their employers' responsibilities in the work-life balance. In addition, companies are regarding their social responsibilities as part and parcel of being a good employer, both in order to attract and retain employees. Some companies regard the composition of their workforce as linking their social responsibility with their market orientation.

Business customers are increasingly imposing supply chain assurance and auditing systems, particularly international branded businesses which are in turn responding to social regulation of western NGOs (see below). Moreover, competitors can also be a driver of greater business responsibility as they use their social involvement as a feature of their competitive branding.

Turning to social regulation, NGOs have emerged as prominent shapers of social agendas which articulate social expectations of business. NGOs such as Greenpeace, the World Wildlife Fund, Amnesty International, Oxfam have developed critiques of individual businesses and types of business practice rather than just of governments and capitalism in general. Whereas initially these NGOs tended to take an adversarial perspective on corporations, there are now instances of more cooperative relationships such as Amnesty International's collaboration with the International Business Leaders' Forum in developing a Human rights road map. The impact of NGOs on business responsibility agendas has been assisted by IT developments enabling ready communications between developing world and western NGOs and by the interest of the western media in bringing NGO concerns to wider public attention. Thus, issues such as the working conditions in developing country suppliers of western countries have become familiar with wide sections of western societies.

Governments themselves have also taken an interest in encouraging increased business responsibility. Although some of definitions of corporate social responsibility would appear to exclude activities that are required by law or regulation, many governments have sought to use various forms of soft regulation to encourage business to take greater responsibility for social agendas through mandating, partnering, facilitating and endorsing (Moon 2004). The Australian Prime Minister's Business Leaders' Roundtable and the UK Minister for Corporate Social Responsibility illustrate governmental interest in endorsing greater corporate social responsibility. OECD governments have sought to facilitate multinational corporations to comply with the OECD Guidelines for Multinational Enterprises by acting as a national contact point to support companies seeking to conform to the standards set out in the Guidelines. The UK government's Ethical Trade Initiative and the CSR Academy illustrate the readiness of government departments to bring their fiscal and organisational resources to partnerships with business and non-governmental organisations in order to advance social agendas in business.

2.3 Corporate Power and Responsibility

Some of the key issues that arise concerning corporations and citizenship derive from both their relationship to other sources of power and from the significance of their power relative to that of others in society. As in other institutions such as governments, churches and trade unions, the power of corporations is itself a resource for irresponsibility, corruption and deception and thus there has been an interest in finding appropriate balances between enabling corporations to fulfil their claimed purposes of meeting demands, employing people and returning profits to owners with restraining them from exploiting the powers that go with these purposes inappropriately. This has been an abiding theme in debates about corporations and society from Charles Dickens' stories of early nineteenth century British capitalism through to more recent debates about the allocation of responsibilities for the Enron and Parmalat collapses and measures to prevent repetitions.

Hence, we take the view that consideration of corporations and citizenship should be contextualised by the themes of power and responsibility. After all, the whole significance of the broader concept of citizenship: it is about identifying, allocating, delineating, restraining, relating and operationalising power and responsibility. Thus political debates have raged about who is or who should be a citizen because of the opportunities that political power affords and the responsibilities that citizens either expect to be shown to them or which are expected of them. Corporations, like people, both have power and are subject to power. They are both attributed responsibility and they claim responsibility. Clearly, power and responsibility are closely related. The possession of power is often a pre-requisite to the ability to take responsibility, yet its possession is also regarded as a reason for which its custodians, users and beneficiaries are expected to exercise responsibly.

Most debates about corporations and power revolve around evaluations of corporations' own power and estimations of appropriate constraints upon them that can be affected by the application of governmental including judicial power (i.e. regulation) or mutual power (i.e. self-regulation). These impositions of power on corporations are often designed to protect investors, employees and societies from the abuse of corporate power. But they also extend to meeting collective business (and arguably societal) interests of enabling fair and free competition among corporations. Debates about corporations and responsibility also revolve around the relative responsibilities that corporations owe to their owners and to their other stakeholders such as their investors, employees and customers, and wider societal interests. This introduces powers afforded by systems of corporate governance. Debates persist here, particularly over who should have power over the corporation and to whom are its executives responsible. This is most vividly illustrated in the somewhat caricatured attribution to Milton Friedman that the responsibilities of managers are solely to the company owners[4] which is pitted against the various applications of stakeholder theory to corporations' responsibilities.

The concept of citizenship enables examination of the ways in which corporations deploy or temper their power to exercise responsibility, to who and why. This can be applied in three ways, first in the ways in which corporations can be considered as citizens. Secondly, certain new roles of corporations are akin to those of governments and therefore raise the question of citizenship rights of people who are affected by corporate activities. Thirdly, and relatedly, following the logic of stakeholder power and arguments about corporate responsibilities to their stakeholders, corporations create arenas for stakeholders to act as citizens, both in respect to the corporations themselves but also in wider societal governance.

It could be argued that our threefold distinctions are rather artificial. We would concede that, from an Aristotelian perspective these distinctions might seem otiose: the three perspectives could be regarded as mutually reinforcing facets of citizenship. However, for the purposes of evaluating

---

[4] He specified that this should be within customary ethics and the law, and also acknowledged the mutual benefits of corporate community investment even though he thought that this should be better described as corporate self-interest rather than responsibility (Friedman 1970).

corporations this approach brings the advantages of general conceptual clarity in a field where this is sometimes lacking, and of underlining the political significance of our dimensions of corporate citizenship. By distinguishing the different power relations and responsibility roles that corporations adopt, we are better able to identify the dynamic qualities of corporations in context. As a result, our findings can be addressed to wider questions of institutional review in global governance. Of course, the three general conceptualisations of corporations and citizenship that we adumbrate are differentially experienced according to the respective societal and corporate governance systems that different political communities have developed.

3 Applying Citizenship to Corporations

We argue that the concept of citizenship is appropriate for consideration of the power and responsibility of corporations for several particular reasons. First, the very fact that corporations use the term corporate citizenship as one of several synonyms for their greater social responsibility warrants taking seriously. This enables us to evaluate corporations in part on their own terms. Secondly, citizenship is a concept which is expressly concerned with social relations of power and responsibility which, as we have suggested, enframe many of the current debates about contemporary business-society relations. More specifically, citizenship is an organising principle for aligning powers and responsibilities *among* members of political communities (i.e. on a horizontal dimension), and *between* them and other institutions wielding power and responsibility (i.e. on a vertical dimension).

Thirdly, the concept of citizenship is at the heart of wider debates about societal governance of which corporations form a key part. Thus, critiques of corporate power are often underpinned by a view that citizenship autonomy and choice are being structured by corporate agendas. Alternatively, there is the view that these citizenship pre-requisites are being undermined as the key institutional representatives of citizens, democratic governments, are being superseded by corporate power. Yet more broadly, there is concern that the contemporary forces of globalisation and the undermining of national governments are also inimical to effective citizenship. Although this latter point does not necessarily directly relate to corporations, by virtue of their role as agents of globalisation (e.g. through foreign direct investment, global supply chains) they are implicated in broader political debates about citizenship. Paradoxically, this point parallels other broader citizenship themes as globalisation raises questions of changing and even multiple citizenship through new patterns of migration and political identity. Perversely, perhaps, the view that governments are becoming increasingly ineffective, be it because either of globalisation or corporations, is also associated with the view that citizenship is endangered by the evidence of voter apathy in many developed political systems (though not, it seems in places where democracy is relatively new such as South Africa, Ukraine, Iraq).

We adopt T.H. Marshall's definition of citizenship as comprising three types of rights: civil, political and social (1965). However, we adapt his classification from simply being rights based and follow the Aristotelian assumption about duties of citizenship, to each other and to the polity as a whole. Civil rights consist of those rights that provide freedom from abuses and interference by third parties (most notably governments), among the most important of which are the rights to own property, exercise freedom of speech, and engage in "free" markets. We refer to these rights and corresponding duties as citizenship *status*. In contrast to these more passive rights (which government respects or actively facilitates) the second category of political rights moves beyond the mere protection of the individual's private sphere and towards his or her active participation in society. This includes the right to vote or the right to hold office and, generally speaking, entitles the individual to take part in the process of collective will formation in the public sphere. We refer to these rights and duties as citizenship *processes*. Thirdly, Marshall's social rights consist of those rights that provide the individual with the freedom to participate in society, such as the right to education, healthcare, or welfare. We refer to these rights and duties as citizenship *entitlements*. In the next section, which fleshes out our three dimensions of citizenship, we will outline different configurations of status, processes and entitlements of citizenship.

4   Three Conceptions of Corporations and Citizenship

As the changing roles of corporations in business-society relations are complex and multi-faceted, rather than cram all of these relationships into a single conception of citizenship, we present three different ways in which the concept of citizenship illumines the powers and responsibilities inherent in business-society relations. In each of these conceptions, we distinguish different roles and relations for *corporations*, for *governments* and for *citizens*, by which we also refer to what others describe as the third sector, or societal NGOs.

4.1   Corporations as Citizens (see Figure 1)

The first conception focuses on corporations as citizens that are ruled but also participate in the functioning of the overall political community. Thus, there are ways in which corporations, like other citizens in democracies are members of communities and engage with other members to enhance the social fabric. In addition, like other citizens, corporations periodically bring their interests and values to the formal governmental processes of law making, implementation and adjudication within their political community. In this conception corporations are on a similar horizontal relationship with other corporate citizens and human citizens. Like human citizens, corporate citizens are also a vertical relationship of power with government in which the citizens 'author' the authority of government, most obviously through elections, and thus governments are responsible to these citizens. However, within the parameters of their legitimate authority, governments are also empowered to govern all citizens. Corporations can be considered as if they were citizens in as much as they work 'with' and participate 'in' society and in bringing their concerns to government and reacting to government legislation and executive action. The focus here, then, is on how corporations share status and process elements of citizenship.

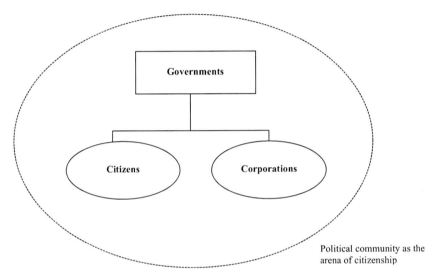

Figure 1: Corporations as citizens

Arguments about corporations being like citizens can have a number of sources, though these are not necessarily mutually consistent. Since medieval times, European business people engaged in citizenship-like ways through their membership of and participation in their guilds, the forerunners of business associations, which provided systems of governance within individual trades and forms of mutual support. In early phases of representative politics, business ownership enabled people to meet a property franchise. Secondly, corporations could be considered as part of society in that their members, be they owners, managers or employees are human members of societies. Relatedly, traditions of industrial paternalism or industrial philanthropy speak of the social face of business. Theories of business legitimacy are often premised on the need for corporations to win the approval of society for their individual and collective existence and success. Some theories identified social objectives with normal business activity:

> Building a better community; improved education; better understanding of the free enterprise system; an effective attack on heart ailments, emphysema, alcoholism, hard chancre or other crippling disease; participation in the political party

of choice; and renewed emphasis on regular religious observances are examples of such further goals. (Galbraith 1974:184)

The idea that corporations could be considered as if they were citizens can also draw on the slightly different sort of argument, that corporations have a distinct functional identity: they are praised or blamed, they make deals, enter into contracts and develop internal decision-making system and structures independent of the people within the company. A further variant is the significance of corporations' distinct legal identity. In essence, incorporation presumes that the businesses is recognised as being capable of acting il/legally and as having duties and rights of legal protection and compensation. Businesses can enter into legal agreements, own property, employ workers, sue and be sued. As a result a company can be treated in the eyes of the law as if it is an 'artificial person'. More generally, the application of the citizen metaphor to corporations can draw on the argument that "every large corporation should be thought of as a *social enterprise;* that is, as an entity whose existence and decisions can be justified only insofar as they serve public or social purposes". (Dahl 1972:17)

Although there are clearly limits to the application of the citizenship metaphor to corporations particularly regarding their *status* (e.g. they do not vote or sit on juries), nonetheless they participate in various *processes* of citizenship. First, corporations engage in various forms of lobbying, be it of governments or of business associations or of the media. This is akin to pressure group activity, justified in liberal democratic politics as an extension of participation through voting. Secondly, corporations participate within community processes of decision-making and mobilization. This might include membership of 'social' partnerships with non-profit and governmental organizations. These might be concerned with such matters as local economic development, education or environmental concerns. Thirdly, corporations can align their activities with broader social agendas as captured in the terms sustainability and 'triple bottom line' thinking, with its commitments to social justice, environmental responsibility, and economic development (Elkington 1999). Corporations may even open their own processes to social engagement as in systems of stakeholder reporting and in deliberation over the targets of corporations social investments. Moreover, corporations can enjoy *entitlements* which are akin to those of citizens such as protection under the law and eligibility for subsidies under various public policy regimes (e.g. for training programmes).

4.2 Corporations as Governments (see Figure 2)

Here we refer to the ways in which corporations are acting as if they were governments and are responsible for the delivery of public goods and for the allocation, definition and administration of rights. This could either be in substitution for government, in the absence of government, or in areas beyond the reach of governments, specifically internationally. Such developments raise important questions for the governing of citizenship even though the cases of corporations replacing citizens entirely are rare (e.g. 'company towns', corporations' health and education systems in developing countries). In such a conception the corporation shares a horizontal dimension with government and is vertically aligned with human citizens within a political community. The focus here, then, is how corporations inform the status, processes and entitlements of people as citizens.

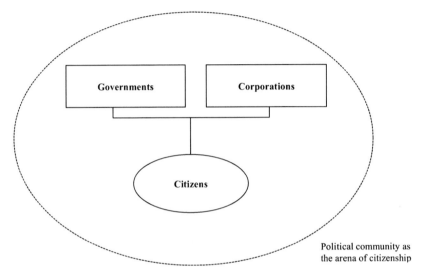

Political community as the arena of citizenship

Figure 2: Corporations as governments

First, corporations might become involved in governing citizenship where *government ceases to do so*. This situation mostly occurs as a result of institutional failure and new political ideology in liberal democracies, and in the shift from communist to capitalist systems in transitional economies. This may happen either when corporations have opportunities to step in where once only governments acted or where corporations are already active but their role becomes more pronounced if governments retreat.

Thus corporations can become more responsible for citizen *entitlements* of former public services. We see corporations increasingly active in the takeover of former public services, such as public transport, postal services, healthcare or education. In fact, many so-called 'corporate citizenship' initiatives are fundamentally equivalent to corporate philanthropy and targeted at reinvigorating (or replacing) the welfare state, such as improving deprived schools and neighbourhoods, sponsoring university education or the arts, or setting up foundations for health research.

Although the *status* of citizens is generally the preserve of governments in developed countries, corporations become directly involved in the ways citizens can claim status by their participation in labour and product markets and in down-sizing industries where governments may have taken responsibility. Similarly, civil rights of prisoners are increasingly a corporate responsibility as correctional and security services are privatized. Governmental failures in developing or transforming countries shift the focus to corporations as Shell found in Nigeria when it was implicated in the failure of the state to maintain the protection of the civil rights of the Ogoni people. It is now suggested that corporations should 'step in' when the status of citizens is threatened in such circumstances.

In terms of *citizenship processes*, corporate roles appear more indirect in that they can help facilitate, enable, or block certain political processes in society, rather than directly taking over formerly governmental prerogatives. Thus many political issues are now directed at corporations rather than at governments (e.g. anti-corporate protests, consumer boycotts). Hence, rather than replacing governments, corporations here could be said to have provided an additional conduit through which citizens could engage in the process of participation.

Second, corporations become active in the citizenship arena where government has not as yet assumed the task of governing. Historically, this was the situation that gave rise to paternalistic employee welfare programmes by wealthy industrialists in the nineteenth century. More recently, exposure to this situation for multinationals is particularly a result of globalization, where lack of local governance in developing countries presents corporations with a choice as to whether to step in as 'surrogate' governments. Corporations such as Nike, Levi Strauss and others which have ensured employees a living wage, and finance the schooling of child laborers have entered into relationships concerning *entitlements* with citizens of developing countries. This possibility may be extended by TRIPPS agreement in which large pharmaceutical companies undertake obligations to provide free or discounted drugs where governments are unable to provide them.

In the case of *citizenship status*, there is evidence that corporations can encourage or discourage oppressive regimes extending citizenship status as under the apartheid period of South Africa and more recently in Burma, Chad, Uganda and Sudan. More widely a new range of civil rights and other status issues are emerging, in particular, issues of privacy and protections of basic freedoms, surrounding new IT and biotechnology industries. These responsibilities often emerge because governments have not worked out their preferred regulatory regime but nevertheless, can have massive implications on life choices of citizens. Similarly, in *processes* of citizen participation, corporations can act as a default option in the absence of government responsibility as in Burma where citizens dispossessed of rights to vote might turn to lobby corporations.

Third, corporations become involved where the governing of citizenship is beyond the reach of the nation state. These situations are a result of the globalization of business activities, an increasing liberalization and deregulation of global economic processes, and escalations in trans-border activity by corporations in which citizen status, entitlements and processes are associated with supranational or *deterritorialized* entities such as global markets or the ozone layer.

Corporations can impact on *entitlements* through their leverage for "favorable" conditions for foreign direct investment which can translate into low social standards, depressed wages, and limited regulation of working conditions. Accordingly, it can become incumbent upon the actions of MNCs to protect (or not protect) social rights, such as through the introduction of global codes of conduct. Due to the globalization of certain financial markets nation states have only limited ability to protect certain aspects of *citizenship status*, particularly property rights over pensions and insurance.

Current changes in global governance have given impetus to corporations' role in governing *processes of citizenship* particularly with self-regulation through programs such as the Chemical Industry's Responsible Care or the Apparel Industry Partnership. Also corporations are playing an increasingly prominent role in such global regulatory bodies as the WTO, GATT or the OECD that have significant impacts on the way governments all over the world govern their relations with their citizens.

4.3 Stakeholders as Citizens (see Figure 3)

Our third conception of citizenship introduces a rather different perspective upon corporations as it envisages circumstances whereby corporate activity itself can shape opportunities for corporations' stakeholders to act as if they were citizens in relation to the corporation. In this conception, corporations are aligned in vertical relations with a variety of stakeholders in the context not of governing the political community (as in our first and second conceptions) but of the corporation (or, of corporate governance). The focus here, then, is on how corporations constitute an arena in which people can engage in citizenship processes, which may include engagement concerning the definitions of their status and entitlements.

Clearly the ways in which stakeholders' citizenship status, entitlements and processes elate to corporations varies enormously among individual stakeholder types such as investors (or owners), employees, customers and societal groups, and among national business systems, but generally the issue of rights has been central to stakeholder relations both in the normative (Donaldson and Preston 1995) and strategic variants (Freeman 1984).

Although the ownership relationship of investors to corporations is at one level a simple economic one, it does also raise issues of power and accountability which are not unlike certain citizenship issues. This political

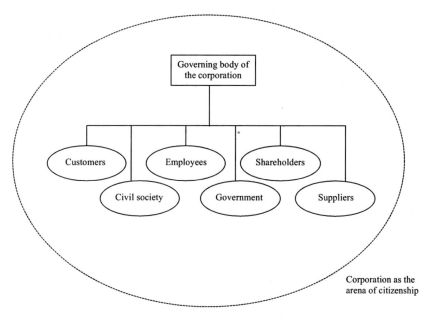

Figure 3: Stakeholders as citizens

dimension is most evident in the notion of *shareholder democracy*, which presumes that a shareholder is entitled to have a say in corporate *processes* rather than simply accepting blindly managers' decisions (e.g. over executive pay, board membership). Social responsible investment funds now increasingly engage in extended dialogue with corporate leaders over issues such as human rights, diversity and labour standards

Employees are the stakeholders that are most frequently conceptualised in citizenship terms (Organ 1988) even though the usage has tended to emphasise solidarity rather than rights and duties. Even the "Harvard Business Review" countenanced the idea of 'building a company of citizens' through the Athenian model of citizenship as a new democratic model of management (Manville and Ober 2003). However, employees also enact processes of participation through engagement with financial (through shareholding) and operational engagement, ranging from the most explicit in cases of negotiations about down-sizing to the, usually, more humdrum in the implementation of regulation and self-regulation. These *processes* clearly vary among national business systems but, again depending on those systems, these can also go to the heart of employee *status* and *entitlements*. In some cases we have seen evidence that corporations encourage some aspects of employee citizenship (e.g. solidarity with the firm) but at the same time discourage other aspects (e.g. *status* of unions).

The idea of consumer sovereignty, central to justifications for markets, literally embraces the language of citizenship, reflecting freedom and authority. Although this is conventionally associated with the quality, price, and availability of product offerings, 'ethical' or 'political' values of consumers have also featured not only in purchasing decisions but also in the mobilisation of these values through NGOs which engage in more direct citizenship *processes* with corporations. In the cases of some aspirational goods, consumers can also acquire some citizenship-like status of certain branded corporations through their solidarity with the product/brand (e.g. Harley-Davidson). Clearly, the relationships between 'consumer' and 'citizen' role are difficult to disentangle, especially when citizens are increasingly being encouraged by governments to behave like consumers.

The combination of corporate global power and expectations of responsibility have brought the supply chain into new sharp relief as part of the arena of corporate governance. The power to purchase brings responsibilities to suppliers especially where purchasers are moving away from short-term, adversarial relationships. This can afford some suppliers, most obviously in the Japanese model, greater 'insider' status and with it the informal, partial, and voluntaristic nature of partnerships which may enable protection of *status* even though this might inhibit wider *processes* of supplier 'democracy'.

Finally, civil society organisations have often been corporations' greatest critics in purporting to represent the interests of sections of society, society in the round and even the environment. Thus they may sometimes resemble human citizens at their most disgruntled. More broadly, they have been welcomed as an adjunct to formal modes of political citizenship because they offer avenues for self-development, active involvement in the community, as well as a form of collective representation to, or resistance to, government and other powerful actors through associations. On the other hand, they are the least formally engaged of the stakeholders we have considered, which inhibits the clarification of their *status, process* and *entitlement* relations with corporations.

5 Conclusions: Citizenship and Corporate Responsibility for Innovation

Through consideration of the different citizenship relations of corporations we have attempted to signal how a host of powers and responsibilities which corporations have acquired or are attributed contribute to a full understanding of the social and political underpinnings of their market operations. Rather than see corporate social responsibility, corporate power and corporate stakeholders as entirely distinct topics, as they are often treated in the literature, we have seen them as reflecting different aspects of these power and responsibility relations. Moreover, all three perspectives have illustrated how roles of corporations do not reflect only their economic operations but also their social and political context. Changes in systems of societal governance and in social demands and expectations have clearly informed the development of corporate roles, for better or for worse. These roles can bring different citizen status, process and entitlements for corporations, citizens, and business stakeholders.

With regard to the topic of this volume the proposed framework actually opens up a couple of implications not only for better understanding public perceptions of corporate responsibility for innovation but also for developing approaches to tackle the issues.

- Thinking of corporations as citizens, they are participating in societal governance chiefly by providing and using certain technological options. The use of stem cell research, genetically modified organisms or genetic data for insurance purposes has severe implications on entitlements of fellow citizens. Key solutions from a citizenship perspective lie in the increased provision of arenas of deliberation with other citizens and offering degrees of participation to other citizens concerned by the technologies (Renn et al. 1995). With regard to the development of new drugs for neglected diseases (mostly of the global poor) a recent study (LSE and Wellcome Trust 2005) suggest that a key solution to approach this contested topic is the use of public-private-partnerships in order to develop these drugs. In a citizenship perspective this does not only reflect the participative nature of the governance of (global) health politics but it also highlights the necessity of protecting corporate entitlements, in particular the entitlements of shareholders on securing viable economic returns while at the same time assuming responsibility for major societal concerns.
- The perspective of *corporations as governments* exposes in particular those situations where corporate innovation has put companies in a situation where they directly or indirectly can either protect or violate fundamental citizenship rights of individuals. This gets immediately visible in the case of big pharmaceutical companies who, unlike governments, actually dispose of theoretical option of providing drugs for treating diseases such as HIV/AIDS, malaria or yellow fever. The companies in fact have a fundamental impact on how of basic social, political and civil rights of individuals are shaped. It would occur then that companies discharge accountability in a similar way governments discharge accountability to their citizens. This would in particular include accountability about their actual cost-benefit situation with regard to the provision of these drugs or providing transparency about the way they shape global trade agreements and influence political decisions pertinent to their business interests. This becomes particularly evident in cases where governments do not protect the status and entitlements of their citizens. In the Chinese example cited earlier then a company such as Yahoo, rather than becoming complicit in infringing fundamental citizen entitlements such as freedom of speech and expression would assume responsibility in protecting these rights through its policies.

- Corporations might be exposed to treating their *stakeholders as citizens* in situations where their stakeholder relations put them in a position to either further or infringe these rights. This, for instance, would suggest to assume a fairly far reaching level of responsibility for the way children use their services. Rather than just treating them as customers, the company would also pay particular attention to the way broader rights of children need to be protected when surfing the web or using chat rooms. On a more general level then this perspective exposes the political nature of corporate decisions on innovation with regard to their stakeholders and suggest polity-like modes of interaction with the stakeholders. A fairly common examples are codes of conduct (Bondy et al. 2006) which companies increasingly use in making their ethical stance explicit and transparent to their stakeholders. A citizenship perspective then exposes these codes more as a political tool similar to laws and ordinances in the political sphere rather than treating them as simple management tools.

Our contribution is thus far mainly conceptual and suggestive of a research agenda which would, first, encourage greater focus on the political aspects of business-society relations alongside the economic. Secondly, it would encourage analysis which considers both power and responsibilities which attend any particular business-society relationship. Thirdly, our distinction of the ways in which business-society relations structure and reflect different citizenship status, processes and entitlements offers a ready framework for research. Fourthly, our approach brings with it normative considerations, particularly concerning the appropriate balances of powers and responsibilities for corporations and other economic, social and political actors.

Bibliography

Albert M (1991) Capitalisme contre capitalisme. Paris: LeSeuil
Attaran A (2004) How do patents and economic policies affect access to essential medicines in developing countries? Health Affairs 23(3):155–166
Barton JH (2004) TRIPS and the global pharmaceutical market. Health Affairs 23(3):146–154
BBC (2005a) Safety urged for child web users. www.bbc.co.uk/news, 18 April 2005
BBC (2005b) Yahoo 'helped jail China writer'. www.bbc.co.uk/news, 9 July 2005
Bendell J (2000) Civil regulation: A new form of democratic governance for the global economy? In: Terms for endearment: Business, NGOs and sustainable development, Jem Bendell, Ed. Sheffield: Greenleaf
Bondy K, Matten D, Moon J (2006) Codes of conduct as a tool for sustainable governance in multinational corporations. In: Benn S, Dunphy D (eds) Corporate governance and sustainability – Challenges for theory and practice. London: Routledge, 165–185
Crane A, Matten D, Moon J (2004) Stakeholders as citizens? Rethinking rights, participation, and democracy. Journal of Business Ethics 53(1):107–122
Crane A, Matten D, Moon J (forthcoming) Corporations and Citizenship. Cambridge: Cambridge University Press
Crane A, Matten D (2007) Business ethics. Managing corporate citizenship and sustainability in the age of globalization. 2nd ed., Oxford: Oxford University Press
Crook C (2005)'The good company. The Economist 22.I.2005, 9
Dahl R (1972) A Prelude to Corporate Reform. Business and Society Review, 17–21
di Norcia V (1994) Ethics, technology development, and innovation. Business Ethics Quarterly 4(3):235–252
Donaldson T, Preston LE (1995) The stakeholder theory of the corporation: concepts, evidence, and implications. Academy of Management Review 20(1):65–91
Elkington J (1999) Cannibals with Forks: Triple Bottom Line of 21st Century Business. London: Capstone Publishing
Freeman RE (1984) Strategic management. A stakeholder approach. Boston: Pitman
Friedman M (1970) The Social Responsibility of Business is to make Profits. New York Times Magazine, September 13
Galbraith JK (1974) The new industrial state (2nd ed.). Harmondsworth: Penguin
Gallie WE (1956) Essentially Contested Concepts. Proceedings of the Aristotelian Society 56(2):187–198
Harris J (1994) Biotechnology, friend or foe? Ethics and controls. In: Dyson A, Harris J (eds) Ethics and Biotechnology. London: Routledge, 216–229
Häyry M (1994) Categorical objections to genetic engineering – A critique. In: Dyson A, Harris J (eds) Ethics and biotechnology. London: Routledge, 202–271
Hertz N (2001) The Silent Takeover. London: Heinemann
Korten DC (2001) When corporations rule the world, 2nd ed. Bloomfield, CT
Langford D (ed) (2000) Internet ethics. New York: St. Martin's Press
LSE, Wellcome Trust (2005) The new landscape of neglected disease drug development (A report from the Pharmaceutical R&D Policy Project). London: London School of Economics/Health and Social Care
Manville B, Ober J (2003) Beyond empowerment: building a company of citizens. Harvard Business Review (January), 48–53
Marshall TH (1965) Class, Citizenship and Social Development. New York: Anchor Books
Matten D, Crane A (2005) Corporate Citizenship: Toward an Extended Theoretical Conceptualisation. Academy of Management Review 30(1)166–179
Monbiot G (2000) Captive State: The Corporate Takeover of Britain. London: Pan
Moon J (1995) The Firm as Citizen: Corporate Responsibility in Australia. Australian Journal of Political Science 30(1):1–17

Moon J (2002) Business Social Responsibility and New Governance. Government and Opposition 37(3):385–408

Moon J (2004) Government as a Driver of CSR' ICCSR Working Papers No. 20 Nottingham: Nottingham University Business School

Moon J, Crane A, Matten D (2005) Can corporations be citizens? Corporate citizenship as a metaphor for business participation in society. Business Ethics Quarterly 15(3):429–453

Organ DW (1988) Organizational citizenship behavior: the good soldier syndrome. Lexington, Mass.: Lexington Books

Renn O, Webler T, Wiedemann PM (eds) (1995). Fairness and competence in citizen participation. Dordrecht: Kluwer

Taunton-Rigby A (2001) Bioethics: the new frontier. Business and Society Review, 106(2):127–142

Whitley R (1999) Divergent capitalisms. The social structuring and change of business systems. Oxford: Oxford University Press

Wolf M (2002) Response to "Confronting the Critics". New Academy Review 1(1):62–65

Zadek S (2001) The civil corporation: the new economy of corporate citizenship. London: Earthscan

Access to Medicines and the Innovation Dilemma – Can Pharmaceutical Multinationals be Good Corporate Citizens?

Andreas Seiter

Introduction

Pharmaceuticals are a key input in health care systems. For many diseases, they provide the basis of treatment or even the entire treatment. In OECD countries, pharmaceuticals account on average for 10–15% of total health expenditure[1]. In developing countries, the share of pharmaceuticals is usually higher (up to 50% and more), partly because there is only a limited range of other, more sophisticated services available.

Patients in developing countries rate the performance of health care systems to a large extent based on the availability of pharmaceuticals. Attendance rates in African clinics fluctuate based on the availability of drugs – people show up for treatment if they know that there are supplies, and they stay away when they hear from others that the clinic has run out of drugs.

Unfortunately, in today's world we are far away from being able to secure access to safe and affordable medicines for all people on this globe. This is partially a problem of resources – there is not enough money to buy drugs in many poor countries. But even if the resource gap could be closed by increasing wealth or external donor funding, there are additional hurdles that prevent access for poor people, such as corruption and fraud, distribution problems, lack of regulatory oversight leading to counterfeit and substandard drugs being marketed as well as cultural and knowledge barriers.

Institutions like the World Bank work with governments to address the systemic weaknesses in many countries. They provide funding not only for the procurement of drugs, but also for system improvements, training and capacity building. However, one characteristic of the pharmaceutical sector is that most of the activities are private for profit. Not only are all major global or regional manufacturers privately owned and run as for-profit businesses, also the trade and distribution is largely in private hands. Most development country governments run their own public delivery systems which

---

[1] OECD Health Data, 2005, http://www.oecd.org.

Access Framework

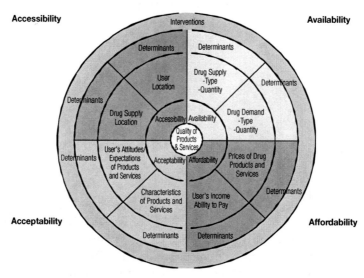

Figure 1: Determinants of access to medicines (Management Sciences for Health)

include procurement of pharmaceuticals, storage and supply to government owned health centers. However, these centers usually cover only a part of the population. Large parts get their treatment in a private sector that has developed its own independent supply systems, in many cases without proper regulation and oversight (see Table 1).

From a World Bank perspective, the question is how this vibrant private sector with its different tiers can be effectively utilized, so that it supports the overall public health goals in addition to pursuing economic interests. What

Table 1: Differences in governance and capacity parameters between the tiers of the private sector supply chain in developing countries

	Integration/ concentration	Regulatory Oversight	Public Scrutiny	Financial and technical capacity
Multinational manufacturers	High	High	Medium	High
National manufacturers	Low	Medium	Low	Medium
Importers/agents	Low	Low	Low	Low
Wholesalers	Low	Low	Low	Low
Retail pharmacists	Low	Low	Low	Low

becomes clear here is that the multinational pharmaceutical companies represent only one of these tiers. While they are key for the development of new drugs and the supply with modern medicines that are not yet widely available from generic drug makers, they are not much involved in the supply of older essential medicines or in the local distribution chain in developing countries. Even if these large companies would completely forego their profit motive and deliver drugs at cost to low income countries, the impact on access to medicines would be limited in absence of a strategy to secure performance and integrity of the entire distribution chain down to the retail level.

Business Model of the Pharmaceutical Industry – Innovative and Generic

The pharmaceutical industry can be divided into two blocks: innovative, research based companies that invent new treatments and sell them at a premium as long as they have patent protection, and generic companies that manufacture well-known drugs and compete over price. Some innovative companies have diversified into the generics business (such as the Swiss Novartis, which owns the generic drug maker Sandoz, currently the second largest globally operating generics company). Some of the larger generics companies are increasingly investing into R&D, trying to win a share of the more profitable business with patented drugs. In addition, the lines are blurred by innovative companies selling their old brands in competition with generic manufacturers, trying to optimize profits by lowering prices in markets that are price sensitive, while maintaining higher prices in other markets. On the other side, some generic drug makers sell their products under brand names and with a premium similar to innovative drugs, predominantly in smaller markets with low levels of transparency and competitive pressure.

What sets pharmaceuticals apart from other products is that the costs for research and development (R&D) are very high, whereas the unit costs for manufacturing are usually low in comparison. In a way, modern pharmaceuticals can be compared with software products that require high investments into development, but can be easily copied once they are on the market. Innovator companies have to refinance the R&D costs during the remaining patent life of a product after launch, which is usually about 10 years (50% of the total patent protection period of 20 years). The price of a product has to be set high enough to pay for (1) the R&D cost of this product, (2) a share of the R&D costs of other products that failed and could not be marketed, (3) the variable costs of the product such as manufacturing, marketing and sales, (4) a share of the company overheads and (5) a profit for the shareholders.

Generic drug companies usually enter the market with a new product immediately after the patent of the originator expires. Large markets such as the US or Germany are targeted first. The price of the first generic is about 20–

30% lower than the original, which is enough to generate significant sales with relatively limited marketing expenses (the product is already well known, customers are waiting for the first generic because they want to utilize the savings potential). At such a price level, a generic product is usually very profitable, as the manufacturer does not have to amortize large R&D expenses. This explains why in competitive and prize sensitive markets generic prices keep dropping until they reach a trough which is defined by the manufacturing costs plus general overheads and a profit margin. This can be 10% of the original price or even lower. At this stage, manufacturers that can make large volumes have an advantage: economies of scale drive manufacturing costs further down – one of the reasons why the Indian pharmaceutical industry is so competitive even compared to other countries with similarly low labor costs.

Both innovative and generics manufacturers, assuming that they do business in developed countries, must have strong capacities in general management, marketing, manufacturing and quality assurance. The pharmaceutical industry in developed countries is highly regulated and under significant competitive pressure. Neglecting quality standards or mismanagement in economic terms can lead to serious consequences including loss of shareholder confidence and takeover by another company. For R&D based companies, the shareholder value is defined by the "pipeline" of new drugs in development and the expected revenue growth based on this pipeline. Many companies that were household names a few years ago meanwhile lost their independence and were "digested" in a merger or takeover after falling behind in a less forgiving market environment. These dynamics must be taken into consideration when thinking of pharmaceutical companies as potential partners in development. On the one hand, a good corporate image is important to create a positive business climate and secure customer loyalty in developed markets. On the other hand, companies will only have limited resources available for development related projects that do not directly contribute to their present or future bottom line.

In developing countries, the managerial and quality standards in the national drug industry vary. Several countries do not have the regulatory capacity to enforce strict GMP (Good Manufacturing Practices) guidelines. Economic ties between political and administrative decision makers and the companies sometimes lead to favoritism and double standards. Another problem is lack of capital to invest into modern technology and attract qualified technical personnel for the operations of a state-of-the-art pharmaceutical manufacturing plant. As a result, there is a significant quantity of substandard medicines in circulation in developing countries. WHO, USP (United States Pharmacopeia) and others regularly publish data on substandard and counterfeit drugs[2]. This issue is at the very basis of Corporate

---

[2] Matrix of Drug Quality Reports on USAID-assisted Countries, United States Pharmacopeia, March 2005, http://www.uspdqi.org/pubs/other/GHC-DrugQuality Matrix.pdf.

Social Responsibility – the first obligation for a business should be to make sure that its products are effective and do no harm.

The Innovation Dilemma

If an innovative company is successful in its mission and develops a truly innovative, life saving medicine, it finds itself in moral dilemma between economic logic and the human obligation to help. Being in possession of something that can save people's lives, the company is a target for consumer activists and media who will demand flexibility in pricing and generosity in giving. Business logic, on the other hand, demands that it asks a price as high as possibly viable in the market – excluding people who cannot pay such a high price from access. A price that is accepted in a rich country, where the population is covered by insurance, is most likely beyond affordability in a poorer country, for public buyers, insurers and private payers equally. Even in a rich country, a new treatment for a disease that so far was untreatable creates additional expenses and budget pressures for health systems. On a domestic scale, companies respond to this pressure usually by generating patient assistance programs. Such programs offer free treatment for uninsured people under a certain income level. They are common in the US where many people do not have coverage for drugs under their health insurance.

The battle over patents and prices for antiretroviral drugs against HIV, which climaxed in a lawsuit against the South African government (1998–2001)[3], shows the international dimension of this conflict. Since then, companies have developed various models to offer new essential drugs either at a steep discount[4], or through subsidized treatment programs that include elements of prevention, patient education and system development. Another alternative are out-licensing contracts that allow national manufacturers to make these drugs for local or regional use in countries that cannot afford to pay the price of the original.

Despite several successful models for improving access to important medicines in various developing countries, the conflict between poor countries keen to obtain newly developed essential drugs at the lowest possible cost and companies who want to preserve their profits is continuing, as the debate around the CAFTA agreement (free trade agreement between the US and several Central American countries) demonstrated. It is particularly difficult to find solutions in countries like Brazil that have significant purchasing power in the hands of the upper and middle class, but at the same time a large number of poor people that receive subsidized health benefits from a

---

[3] 2001: Drug firms withdraw from AIDS case, BBC News, On This Day, 19 April 2001, http://news.bbc.co.uk/onthisday/hi/dates/stories/april/19/newsid_2488000/2488509.stm.

[4] Accelerating Access Initiative, http://www.ifpma.org/Health/hiv/health_aai_hiv.aspx.

financially overstretched government or have to pay out of pocket. This conflict between the business side of innovation and society's expectation to get fast and affordable access to new life saving technologies is a significant causal factor for "Big Pharma's" chronic image problem. But the same conflict is also a driver behind many of the "Good Corporate Citizenship" initiatives launched by major drug companies in an attempt to fulfill their perceived moral obligation.

Incentives for Innovation

Given the business model of innovative pharmaceutical companies, it is clear that the focus of their R&D will be on treatments for diseases that affect people in affluent markets. Nevertheless, born out of the dialogue with Global Society, there are an increasing number of voluntary initiatives by major drug companies to utilize their R&D capacity for the benefit of people in the developing countries. In the absence of a significant market, however, it is difficult for a for-profit company to justify major expenses for the final stages of development of a product that targets diseases of people who cannot pay for treatment. Governments of major donor countries are aware of this dilemma and have responded for example by developing a concept for advanced purchasing agreements for vaccines[5]. It would offer a manufacturer an attractive price for a product that fulfils certain specifications. After a certain time, when the manufacturer has sold enough to refinance the R&D investment, the contract foresees a steep price cut that would free donor funds for future innovations. In general, it can be said that in absence of an affluent market, there must be a committed institutional buyer or a financial risk sharing agreement to allow drug companies to make a full investment in medicines for so-called "neglected diseases". For diseases that have at least a limited market in developed countries or countries with a partially affluent population, the company that develops the drug may be able to recover the development costs if it gets an acceptable price and a period of marketing exclusivity for these affluent markets. Orphan drug laws, as they are in place for example in the EU and the US, allow for such additional exclusivity for manufacturers who invest in diseases that affect only a small population.

Public Private Partnerships for Innovation in Areas Neglected by the Market

In recent years, a number of partnerships have been initiated, funded by international donors and large foundations, which focus on specific disease

---

[5] Introducing Advance Purchase Commitments, DFID consultation note, 2005, http://www.bvgh.org/documents/DFIDAPC2-pager.pdf.

areas relevant for developing countries[6]. As far as these partnerships have external funding, they can outsource parts of their R&D needs to pharmaceutical companies or specialized research contractors. In such relationships, the partnering company may offer some pro-bono work, access to proprietary information or charge service fees below market rates, but it does not bear the financial risk of the development process. The progress of these partnerships has been encouraging so far, but the final test will be whether they are resourceful enough to bring a product to market. At the final stages of development, when costs go up exponentially, the donors behind the partnership will find themselves in the same position as described in the previous paragraph: Without a realistic expectation for donor funding to pay for the final product, the investment into late stage development will be hard to justify.

Developing Country Needs in the Pharmaceutical Sector

The public discussion about developing country needs in terms of pharmaceuticals is usually focused on two issues only – price of drugs and innovative treatments for neglected diseases. Part of the problem is that the official voices of these countries frequently, although with notable exceptions, represent a system that caters more to the needs of bureaucrats than of patients and health care workers. Only when a wider range of stakeholders is interviewed on the ground, it is possible to get a more complete picture. As a matter of fact, in those countries that show the worst performance in terms of health indicators, the main problems are lack of access to very basic essential drugs that are available on the global market in good quality at low prices. Due to dysfunctional health systems, these essential drugs are not reaching the people, because of one or several weak links in the distribution chain, starting from poor planning, bad storage conditions, theft from warehouses, inadequate financing at the central or regional level and other factors. In addition, the lack of knowledge and shortage of well trained health workers and physicians lead to suboptimal use of available medicines.

The launch of an ambitious treatment strategy with innovative medicines, for example for AIDS or malaria, requires substantive efforts to address systems related problems in such an environment. This is one of the areas in which the World Bank and other multilateral organizations are engaged and trying to assist countries in upgrading their systems. However, these organizations frequently lack business experience and may miscalculate the dynamics of the private sector. Here is a true chance for better collaboration and new types of partnerships in the future. Sometimes it requires an external initiative such as a treatment program sponsored by a major company or a partnership of donors to create pressure on inadequate systems and catalyze the changes necessary for better performance.

---

[6] World Health Oganization, Public-private Partnerships, http://www.who.int/neglected_diseases/partnership/en/.

Economic Considerations versus Equitable Access

It has become relatively easy to convince multinational, innovative pharmaceutical companies to provide an important medicine at low costs to Least Developed Countries, for example in Sub-Saharan Africa. Due to the poverty in these countries, there is hardly any business opportunity in the public or private sector for highly priced medicines. Therefore the introduction of a low price version of an innovative drug does not cannibalize the company's sales, unless the product finds its way outside the country into more affluent markets. In addition, the "PR value" of corporate programs targeted at African countries is higher compared to programs in other parts of the world.

More problematic is the situation in Middle Income Countries, in particular those with a significant size such as Brazil, India, China, Indonesia and others. For pharmaceutical companies, these markets are commercially important. Given the growing restrictions in terms of pricing and reimbursement in Europe and the growing strength of buyers in the US, companies must look at larger Middle Income Countries for future growth. Although the share of total sales represented by the emerging markets was only around 12% in 2004[7] (IMS Health), they are growing about twice as fast as the mature markets. Therefore, companies are reluctant to offer Middle Income Countries the same low prices as Africa. The economically rational strategy would be to target the high income segment in these countries and charge prices at OECD level. As a consequence, poor people would be unable afford these new treatments, and public or private payers and insurers would have to restrict access based on what they can afford within their limited budgets. This could create significant conflict potential and political pressure on the government. Drug companies are facing tough negotiations when they want to register their drugs in these countries. The usual outcome is that the drug is registered, although with a price significantly below the OECD price level. Nevertheless, this price is usually too high for the poor majority in these countries. Thus the outcome can be characterized as a lose-lose situation: The company cannot fully exploit the economic potential of the wealthy part of the market, while the poor part still has no or only limited access to a new treatment.

The Swiss drug maker Novartis has tried to address this issue by offering a global patient assistance program for its leukemia treatment Glivec®[8]: In all countries that register Glivec and respect its patent, patients who cannot afford to pay the high price become eligible for free treatment. Eligibility cri-

---

[7] IMS Health, Global Insights, World Markets, July 2005: Looking to China and Cancer as Cost Containment Slows Growth, www.imshealth.com/globalinsights.
[8] Glivec® International Patient Assistance Program (GIPAP), https://www.maxaid.org/.

teria are defined based on insurance coverage, personal assets and income. The program is administered by an independent third party; the application has to be made through a physician who is registered with the administrating entity. So far, more than 11,000 patients in 80 countries are covered by this program (outside the US; within the US Novartis runs a separate patient assistance program). However, such an approach is only possible if the number of patients is small, like in the case of CML (chronic myeloid leukemia). For more common diseases, other models need to be developed in order to address the conflict between economically rational behavior of a profit seeking entity and the equity goals of public health policy.

Can the Industry Help to Improve Pharmaceutical Policy?

Pharmaceutical policy defines the rules for manufacturers, distributors and retailers of pharmaceutical products. It sets standards for the administration on how to enforce efficacy, quality and safety of drugs and provides the framework for reimbursement of drug expenditure through insurers, social security systems or direct government budget support. Pharmaceutical companies try to influence policy makers in favor of legislation and administrative guidelines that suit their economic interests. However, the pharmaceutical business model is very capital intensive and long term oriented. Therefore, the big companies have a vital interest to maintain a stable political and economic environment, and to have significant regulatory hurdles in place that prevent lower quality manufacturers from entering the market "on the cheap". Thus, industry interest and general public interest are not necessarily

Table 2: Potential for collaboration between the public sector and multinational pharmaceutical companies in various policy areas

Policy Area	Industry Expertise	Alignment of Goals
Regulatory process	High	Medium
Good Manufacturing Practices	High	High
Forecasting and planning	High	High
Procurement	Medium	Medium
Pharmacovigilance, quality assurance in the supply chain	Medium	High
Pricing	High	Low
Reimbursement	Medium	Low
Rational use of drugs	Medium	Low
Advertising and promotion	High	Low
Consumer education	High	Medium

diverging in all areas. There is a shared interest in efficient regulatory procedures, fairness and transparency, as well as safety and quality hurdles that keep sub-standard products off the market. Table 2 shows in which policy fields industry and public interests are aligned and in which they might be in conflict. In aligned areas, a closer collaboration between policy makers and industry experts would benefit both sides. Even in areas with generally conflicting goals, there might be specific projects which could justify close collaboration, such as a differential pricing strategy for middle income countries with large internal income variations.

What Can Realistically be Expected from the Industry? What are Limiting Factors for the Industry's Participation in the Global Development Effort?

Pharmaceutical multinationals have become more responsive to the needs of developing countries and the development community, partially due to external pressure, partially because they see it as strategic necessity to renew their "license to innovate" in the developed countries. Today, all major companies are engaged in some kind of access program. Many companies provide certain essential drugs to poor countries for free or at cost, others engage in comprehensive prevention and treatment programs (such as Merck in Botswana) or collaborate with national manufacturers who can produce cheaper local versions of important drugs under license. A company that is developing a new antiretroviral drug against AIDS will routinely consult with development organizations or NGOs and come up with an access strategy for poor countries prior to marketing the drug. These company-owned programs may not be satisfactory to address all needs of developing countries (in particular the middle income countries are problematic as explained in the chapter on the "Innovation Dilemma"), but there is a certain standard that will most likely prevail and allow for workable solutions that accommodate some of the needs of developing countries.

There are also a number of relatively recent product development partnerships and initiatives, addressing several "neglected diseases" and providing resources in kind (access to compound libraries and research tools) as well as financing for the development of innovative drugs for diseases of the poor. Typically, the larger corporate partner in such partnerships retains some rights for marketing any successful end products of such joint programs in developed countries, while waiving patent rights and offering technical assistance for registration and launch in developing countries. These partnerships might be harder to sustain or grow if companies come under financial pressure, as long as there is no direct economic incentive to invest into public health related projects.

A potentially limiting factor for voluntary access projects is the difficulty to manage such projects successfully. Everybody in the development community knows how hard it is to implement a successful treatment program in a country with very limited health systems resources and governance structures. Children die every day form dysenteric diseases despite the availability of cheap life-saving oral rehydration products. Malaria deaths are far too frequent and on the rise, despite the availability of effective medicines. The compounded problems of poverty, lack of knowledge, poor infrastructure, lack of human resources and corruption undermine many well intended projects and lead to frustration among donors who expect that their effort will make a difference on the ground.

Pharmaceutical companies do not have a significant infrastructure in poor developing countries. They have to rely on the capacity of local institutions and development organizations, both working in ways that are unfamiliar for someone used to the standards of a large multinational firm. If a project fails due to weak local implementation capacity, the blame might fall on the company that sponsored it as the most publicly visible of the partners. This may have a negative impact on the willingness to engage in future partnerships.

Ideas for further Development, Innovative Partnership Approaches

Recently, there is a significant amount of debate about longer term incentives for the private sector to engage in ways that benefit developing countries. Supported by the Gates Foundation, a model was developed for advance contacting with vaccine manufacturers, offering a guaranteed price and volume for new vaccines needed to fight infectious diseases such as meningitis, Dengue fever and others. There is a good chance that the G-8 governments step in to set up the required guarantee funds to make this model a reality. A similar option could be considered for new drugs.

Alternatively, orphan drug laws could make sense if a disease affects developed country populations as well, but is too small an economic stimulus to ensure its inclusion in the mainstream for profit research and development pipelines. Orphan drug laws offer a defined period of exclusivity for an innovator, independent of existing patent rights.

Another option that has been debated but not been pursued so far are tradable extensions of patent rights in developed markets: A company that makes a defined contribution to public health in developing countries would get a certificate that can be exchanged for an additional 3 months (or other period of time) of patent life in major markets for any product of choice. This certificate could be traded so that the incentive would have the same market value for all companies, whether they have patented products in the market or not.

National legislation could be used to encourage foreign companies to invest into local manufacturers and help them move up to international standards. As pointed out in previous chapters, there is ample opportunity to invite industry to share know-how for regulatory and administrative reform, although conflicts of interest need to be carefully considered. Companies are providing such consulting services already on a bilateral basis, mainly in order to improve the regulatory framework in Middle Income Countries and create a level playing field for foreign and national manufacturers.

Overall, there is no doubt that the private sector can play a significant positive role as a partner for global development strategies. This is not limited to specific sectors such as the pharmaceutical industry. Any major company doing business in sub-Saharan Africa should have good reasons for example to provide basic health services for its workforce, in order to reduce the economic impact of malaria and HIV/AIDS. If they succeed in higher health standards and behavior changes in a part of the population, such activities have positive externalities. Logistics companies could assist in developing better distribution systems for drugs, Cell phone operators and internet café owners could be encouraged to utilize their technology for the dissemination of critical health information and educational messages, for example for AIDS prevention. To realize the potential of the private sector, it does not only require support from the business community. Much more important would be a change of attitude in the donor community and local governments, who need to start thinking "outside the box" and overcome their own hesitations to engage private companies as full partners.

IT Innovations and Open Source: A Question of Business Ethics or Business Model?

Markus Nüttgens

1 Best Practice for Free?

The internet makes data become an increasingly transient good. Almost any song and even entire movies are available in a digital form – not always to the delight of the copyright owners. Another type of good, which can be considered as predestined to be distributed over the channel *internet*, is software. Vendors use the internet for direct sales purposes as well as for making software explicitly available free of charge, what offers interested people the possibility to test preliminary software versions *(beta versions)* and to report possible software errors *(bugs)* to the vendor. This concept of distributed quality assurance isn't very innovative. The idea of Open Source even incorporates the integration of all involved people and all parts of a software product's lifecycle via an open license model (DiBona et al. 1999, Feller et al. 2005).

To put it in a nutshell, the core idea of Open Source is that users do not only participate in the testing process of software but also actively improve the software itself by performing changes on the source code level. To give users the possibility to do so, the source code of Open Source software is often freely available over the internet – this can be considered as the very opposite of common software business models. This paradigm gives the opportunity to incorporate the skills and experiences of software developers all over the world to improve the quality of a software product and to extend it with new features. Several experiences of the last years have shown that such collaborative and open software development processes can decrease the amount of software bugs to a much higher extent compared to non public source code products. It is important to mention that the term Open Source does not mean that the usage of such software produces no costs – the biggest part of the Total cost of Ownership (TCO) is still given by the rollout, user training and support.

The example of the prominent operating system Linux shows that Open Source has finally changed from an idealistic idea to a threat for commercial software providers. One of the currently most successful open software projects is the "Apache" web server that occupies the top position among the most used web servers (Apache 2006). From about 76 million web server installations world wide, in February 2006, the Apache web server had a market share of around 68 percent (Netcraft 2006).

Even governments begin to think about the usage of Open Source software for public service purposes more often. One driver of such thoughts might be the suspicion that sensitive data could be accessible for third parties, if commercial software is used. A famous example that impressively underlines that the publication of a software's source code is no security risk at all is given by the encryption software "Pretty Good Privacy" (PGP) (PGP 2006). In this case, the accessibility of the source code greatly improves the security level of the software by offering third parties the possibility to take an insight into the software to look for possible problems.

These developments won't leave the consulting business unaffected. To give customers strategic advices for the design of information systems and to consult a customer on the operative level during the rollout of solutions, the competence portfolio must be extended accordingly. Thereby the consulting companies decrease their dependence from single software vendors. Above this, the possibility to proactively develop specific solutions arises. If this development is performed together with the customer, a new dimension of customer loyalty emerges.

In the following section the concept of Open Source will be explained shortly (for a detailed introduction see Nüttgens and Tesei 2000). After that, upcoming action options for software companies and consultants are introduced and discussed with respect to the Open Source concept.

2 What is Open Source?

Open Source refers to the free availability of software source code to use and change it according to ones personal needs. This approach at first seems to be the total opposite of classical software sales models that normally protect the software source codes to avoid an uncontrolled spreading of the software.

Currently the business models for Open and Closed Source appear to be contradictory. Figure 1 compares the core aspects of these business models. While traditional business models tend to match the left column, Open Source business models are more likely to match the right one.

In the following, the terms *free software* and *open software* are used synonymously for Open Source software. They express that software can be used, copied and distributed by anyone. This might be done free of charge or not – but in any case the software source code must be made available to every interested person. Open Source software is not always free of charge per definition. Fees can arise for distribution services for example

Feature Type	Feature	
Distribution	Licensed party	Free redistribution
Technical platform	Proprietary	Independent
Program code	Binary code	Source code
Organisation	Company	Community
Capitalisation	Licence fee	Services
Coordination	Central	Democratic
Motivation	Monetary	Idealism
Authoring	Anonymous	Personalisation
Distribution	Commercial Sales	Exchange/ Download

Figure 1: Characteristics of software markets

but generally not for licensing issues. Proprietary software can be considered as the opposite of the Open Source concept. The usage, distribution or modification of proprietary software generally requires an approval by the copyright owner.

2.1 The Genesis of Open Source

Ever since the very beginning of software development in the 1960s and 1970s, programs were shared among development teams for reading, changing or reusing purposes (software sharing communities). In 1982 the companies IBM, HP and DEC released commercial versions of the UNIX operation system for their own hardware. Members of various research teams (from universities or even other companies) were recruited and they worked on commercial software to a greater extent from then on. The simple communication infrastructure started to become increasingly insufficient and so the software sharing communities slowly started to dissolve. A vacuum originated within the software production area that motivated companies to fill it with commercial software. This development resulted in the decrease of so called free software. Companies and research institutions decided so use proprietary software to a greater extent. The technological development supported this trend as operating systems were often directly dependent to hardware so that all processors needed their own proprietary operating system very soon.

For reanimating the cooperative spirit of the software sharing communities, the former MIT employee Richard Stallman initiated the GNU (GNU's not UNIX) project and founded the Free Software Foundation (FSF) (see Stallman 2006; FSF 2006). At that time, the project goal was to develop a free and open UNIX operating system (Müller 1999:17). Although a UNIX kernel was never realized, the huge set of free software systems and development tools from the GNU project enabled the development of Linux in the early 1990s. A student named Linus Torvalds started to work on a free UNIX kernel for personal computers with Intel 80386 microprocessors. His rapid success attracted other developers over the internet who started to support him in his efforts to develop the operating system Linux as the first full functional free UNIX. At this time Linux wasn't noticed by many developers. It took another 5 years until the idea of Linux was broadly perceived. At the end of the 1990s an increasing number of developers began to focus on the internet and on Linux. The World Wide Web (WWW) finally made the internet become a mass medium whereby the amount of potential new developers (metaphorically spoken) exploded. Companies that sold proprietary UNIX (like HP, DEC or IBM) had a quite bad marketing at that time. So Microsoft was able to gain huge market shares for its Windows operating system with aggressive marketing campaigns. By 1994 Linux had reached a solid stability level and had become prominent as a development platform. The available programming languages and tools encouraged developers to start new projects for Linux respectively to port already existing projects to this platform. Because universities and research institutions were very interested in Linux, more and more projects (in the fields of programming languages, databases, graphics or desktop for example) were initiated. As the number of fields that were addressed by free software increased, the whole free software community grew accordingly. This also led to an increasing amount of special interest groups. As a consequence of this upsizing, there were a lot of different opinions of useful definitions of the term "free software" soon. There have always been various different licenses for free software and by the engagement of companies like Netscape, Troll Tech or IBM several new ones were added.

The term Open Source became prominent by the Open Source Initiative (OSI) – a non-profit organization with the goal to introduce the spirit of open source to a broader public. The public success of this term was reflected by many articles about Linux and about the publication of the source code of the Netscape browser that were published in 1997 and 1998. On February 23rd 1998 the Netscape Company finally announced that they will officially use the term Open Source. Soon companies like Corel, Sun Microsystems, IBM, SCO, HP, Oracle, Informix and SAP followed by making similar announcements. Nowadays Open Source and the idea behind it is established and noticed by a broad public. The mentioned characteristics of Open Source software are subsumed in the Open Source Definition (OSI 2006a). As the appellation Open Source is descriptive, it can not be registered as a trade mark. Because of the free software community's need for a reliable identification of open source software, the OSI introduced the certification mark "OSI certified" (OSI 2006b). If software is labeled with this mark, the OSI certifies that the software is distributed under a license that is conformable to the Open Source Definition. A discussion of the currently most common Open Source licenses can be found in St. Laurent (2004). As licenses are no goods, they can not be registered at the U.S. Patent and Trademark Office as well. An OSI certification is performed in two steps:

The OSI publishes a list of licenses that are conformable to the Open Source Definition. If a new license should be added to this list, it has to be submitted to the OSI for a discussion. New licenses are accepted if the subscribers participating in this discussion have no objections.

To mark software with the "OSI certified" mark, the software must be distributed under a license that can be found on the OSI approved licenses list (OSI 2006c).

2.2 Development and Distribution

The Open Source approach is based upon certain development and distribution models that are briefly explained in the following. Until 1997 the development process of free software was practically not documented and solely apparent from known projects like Linux. In his essay "The Cathedral and the Bazaar", Raymond tried to describe factors of a successful open source development process for the first time. According to this essay, the open source development model is also known as the "Bazaar method". In the latest version of his essay, Raymond analyses the way the development of Linux took place and extracts the following rules for Open Source development processes out of it (Raymond 2000):

1. Every good work of software starts by scratching a developer's personal itch.
2. Good programmers know what to write. Great ones know what to rewrite (and reuse).
3. "Plan to throw one away; you will, anyhow." (Fred Brooks, "The Mythical Man-Month", Chapter 11).
4. If you have the right attitude, interesting problems will find you.
5. When you lose interest in a program, your last duty to it is to hand it off to a competent successor.
6. Treating your users as co-developers is your least-hassle route to rapid code improvement and effective debugging.
7. Release early. Release often. And listen to your customers.
8. Given a large enough beta-tester and co-developer base, almost every problem will be characterized quickly and the fix obvious to someone.
9. Smart data structures and dumb code works a lot better than the other way around.

10. If you treat your beta-testers as if they're your most valuable resource, they will respond by becoming your most valuable resource.
11. The next best thing to having good ideas is recognizing good ideas from your users. Sometimes the latter is better.
12. Often, the most striking and innovative solutions come from realizing that your concept of the problem was wrong.
13. "Perfection (in design) is achieved not when there is nothing more to add, but rather when there is nothing more to take away." (Antoine de Saint-Exupéry)
14. Any tool should be useful in the expected way, but a truly great tool lends itself to uses you never expected.
15. Provided the development coordinator has a communications medium at least as good as the internet and knows how to lead without coercion, many heads are inevitably better than one.

On the basis of this development model, a complementary distribution model for open source products was established. Developers provide their software on the internet. Interested users can search for software on their own, test and use it. Also users can get into direct email contact with the developers and thereby participate in the development of the software. Another possibility is given by the distribution of software bundles. Contrary to proprietary software, where customers have to pay license fees, fees for free software only arise for the service of providing. This means that free software that is acquired once can be freely passed down to others. The most common distribution types are (1) the download from the internet, (2) the ordering and shipping of CDs or DVDs and (3) the free give away of CDs or DVDs as attachments to magazines.

3 Strategic Options for Consultancies

Open Source challenges the development of proprietary software and consecutive *classical* consulting services while it simultaneously provides big chances for strategic reorientations. Figure 2 shows a strategic action framework for IT related consulting services in the open source context. The dimensions of this framework are (1) the consulting focus (industry oriented vs. software oriented) and (2) the chosen development approach (Open Source vs. proprietary software). While "industrial orientation" means that the provided services of a consultancy focus on a certain industrial sector (e.g. retail or banking), "software orientation" means that the services are aligned with a specific software application and its feature range. The characteristics of the development approach dimension are self describing.

Each of the four fields of this framework is related to a particular kind of consulting service:

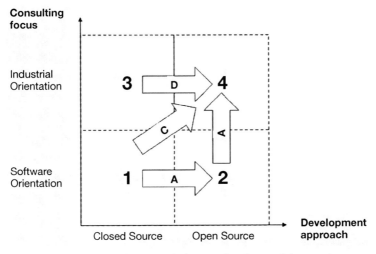

Figure 2: Strategic action framework for IT related consulting services

Field 1
Description: Introduction and customizing of proprietary software systems.
Providers: IT consultancies in the area of office and ERP systems.

Field 2
Description: Introduction and customizing of Open Source systems.
Providers: Companies that emerged from the Open Source community. These companies test, document and distribute Open Source software. Consulting services offered include introduction, user training and technical support.

Field 3
Description: Development of technical-organizational solutions by the usage of proprietary software systems.
Providers: Process consultancies with high industrial sector competences that implement proprietary software systems on the basis of business process analysis.

Field 4
Description: Development of technical-organizational solutions by the usage of Open Source systems.
Providers: Consultancies that develop and implement solutions by using specific Open Source systems that may have been self-developed or extended.

If one reflects the described potentials of Open Source systems and then assesses them superior to proprietary solutions, the conclusion is that positions within field 1 and 3 will lead to competitive disadvantages in the medium-term. The usage of closed software technologies complicates the customer individual configuration and development of solutions.

Consultancies that are positioned in field 2 will hardly participate on profitable projects while it's still possible for IT oriented service providers to exploit this field's potential. Focusing on field 4 can be considered as the most promising strategy because Open Source based solutions of technical-organizational problems offer the greatest benefit for consulted customers. Consequently following the Open Source spirit, consulting services might be openly documented and thereby be made available in a standardized format (as reference models for example). Beyond that, strategic customer-provider alliances can be established on the basis of this open knowledge management. These are advantageous for both parties involved as the cooperative development of detailed problem solutions can be realized. Taken these assessments, transformation strategies can be formulated according to the arrows in figure 2.

Strategy A: Gain technical Open Source competence.

If the core competence is concentrated on the technical level, the expansion into the Open Source sector is a feasible strategy to become a specialized IT service provider.

Strategy B: Gain industry sector competence.

Based upon a distinct technical competence in the Open Source area, the build-up of industry sector competence makes it possible to offer more sophisticated consulting services to the customer – especially with respect to the development of industry specific solutions. In addition, the mobilization of a broad industry sector competence is a good occasion to increase the consulting royalties. If the market share of a consultancy is too small, the development of industry sector competence offers the opportunity of creating unique selling propositions. Building up strategic alliances with other consultancies that are focused on industry sectors might be an alternative.

Strategy C: Gain technical Open Source and industry sector competence.

This strategy is mostly congruent with strategy B. The development of Open Source competence has the higher priority because of the former technology focus.

Strategy D: Gain overall Open Source competence.

The goal of this strategy is to gain a substantial overview of available Open Source solutions to use them target-oriented (and maybe even linked). A lot of technical competence must be developed as well to be able to enhance existing systems. In this strategy, the build-up of strategic alliances with other, technical-focused, consultancies might be an alternative.

In the short- and mid-term, pure Open Source providers also have the possibility to move from field 2 to field 1 to link successful proprietary systems with Open Source systems or to transform proprietary software into Open Source projects. This reveals that the adaptation strategies of the present software vendors must be considered as well.

4 Outlook

The concept of Open Source offers an alternative way for developing and distributing software. It combines already existing knowledge about software development and software distribution and effects of the internet economy. Due to the increasing standardization level of royalty-free software interfaces and formats, the importance of proprietary software solutions is assumed to decrease accordingly. If the service levels are comparable, customers feel more attracted to Open Source distributors to avoid strategic dependencies on vendors of proprietary software solutions. Present software costs will remain existent to some extent, but in the context of Open Source, these costs are no longer software license fees but service royalties. Opened and linked collaborations appear to produce more sophisticated solutions for company problems. The potentials arising offer consultancies the possibility to realign their strategic focuses. Today companies no longer ask the question "Why should we use products whose quality and enhancement is not guaranteed by any company?" but rather "Why should we buy software whose quality is no subject to public discussions?" and "Why should the control of central systems be given into the hands of another company?"

In the field of operating systems, the Open Source concept is already highly established – mainly because of the engagement of distributors like Redhat and SuSE. Time will show if a similar development will take part in the field of application software. The consulting business can play a key role therein.

References

Apache Software Foundation (ed) (2006) The Apache HTTP Server Project. URL: http://httpd.apache.org

DiBona C, Ockman S, Stone M (eds) (1996) Open Sources: Voices from the Open Source Revolution. URL: http://www.oreilly.com/catalog/opensources/book/toc.html

Feller J, Fitzgerald B, Hissam SA, Lakhani KR (eds) (2005) Perspectives on Free and Open Source Software. Cambridge et al., The MIT Press

Free Software Foundation (FSE) (ed) (2006) The Free Software Foundation. URL: http://www.fsf.org

Müller M (1999) Die Philosophie des GNU und die Pragmatik des Open Source. In: Open Source – kurz & gut, Cologne et al., O'Reilly & Associates, Inc., 17–19

Netcraft Ltd. (ed) (2006) Web Server Survey 2006. URL: http://news.netcraft.com/archives/web_server_survey.html

Nüttgens M, Tesei E (2000) Open Source. In: Scheer A-W (ed) Veröffentlichungen des Instituts für Wirtschaftsinformatik Saarbrücken, 2000, No. 156, 157, 158

Open Source Initiative (ed) (2006a) The Open Source Definition. URL: http://www.opensource.org/docs/definition.php

Open Source Initiative (ed) (2006b) Certification Mark and Process. URL: http://www.opensource.org/docs/certification_mark.php

Open Source Initiative (ed) (2006c) Licensing. URL: http://www.opensource.org/licenses/index.html

PGP Corporation (ed) (2006) PGP Source Code Downloads. URL: http://www.pgp.com/downloads/sourcecode/index.html

Raymond ES (ed) (2000) The Cathedral and the Bazaar. URL: http://www.catb.org/~esr/writings/cathedral-bazaar

Stallman R (2006) About the GNU project. URL: http://www.gnu.org/gnu/thegnuproject.html

St. Laurent AM (2004) Understanding Open Source and Free Software Licensing. Beijing et al., O'Reilly Media Inc.

Further volumes of the series Ethics of Science and Technology Assessment
(Wissenschaftsethik und Technikfolgenbeurteilung):

Vol. 1: A. Grunwald (Hrsg.) Rationale Technikfolgenbeurteilung. Konzeption und methodische Grundlagen, 1998

Vol. 2: A. Grunwald, S. Saupe (Hrsg.) Ethik in der Technikgestaltung. Praktische Relevanz und Legitimation, 1999

Vol. 3: H. Harig, C. J. Langenbach (Hrsg.) Neue Materialien für innovative Produkte. Entwicklungstrends und gesellschaftliche Relevanz, 1999

Vol. 4: J. Grin, A. Grunwald (eds) Vision Assessment. Shaping Technology for 21st Century Society, 1999

Vol. 5: C. Streffer et al., Umweltstandards. Kombinierte Expositionen und ihre Auswirkungen auf den Menschen und seine natürliche Umwelt, 2000

Vol. 6: K.-M. Nigge, Life Cycle Assessment of Natural Gas Vehicles. Development and Application of Site-Dependent Impact Indicators, 2000

Vol. 7: C. R. Bartram et al., Humangenetische Diagnostik. Wissenschaftliche Grundlagen und gesellschaftliche Konsequenzen, 2000

Vol. 8: J. P. Beckmann et al., Xenotransplantation von Zellen, Geweben oder Organen. Wissenschaftliche Grundlagen und ethisch-rechtliche Implikationen, 2000

Vol. 9: G. Banse, C. J. Langenbach, P. Machleidt (eds) Towards the Information Society. The Case of Central and Eastern European Countries, 2000

Vol. 10: P. Janich, M. Gutmann, K. Prieß (Hrsg.) Biodiversität. Wissenschaftliche Grundlagen und gesellschaftliche Relevanz, 2001

Vol. 11: M. Decker (ed) Interdisciplinarity in Technology Assessment. Implementation and its Chances and Limits, 2001

Vol. 12: C. J. Langenbach, O. Ulrich (Hrsg.) Elektronische Signaturen. Kulturelle Rahmenbedingungen einer technischen Entwicklung, 2002

Vol. 13: F. Breyer, H. Kliemt, F. Thiele (eds) Rationing in Medicine. Ethical, Legal and Practical Aspects, 2002

Vol. 14: T. Christaller et al. (Hrsg.) Robotik. Perspektiven für menschliches Handeln in der zukünftigen Gesellschaft, 2001

Vol. 15: A. Grunwald, M. Gutmann, E. Neumann-Held (eds) On Human Nature. Anthropological, Biological, and Philosophical Foundations, 2002

Vol. 16: M. Schröder et al. (Hrsg.) Klimavorhersage und Klimavorsorge, 2002

Vol. 17:   C. F. Gethmann, S. Lingner (Hrsg.) Integrative Modellierung zum Globalen Wandel, 2002

Vol. 18:   U. Steger et al., Nachhaltige Entwicklung und Innovation im Energiebereich, 2002

Vol. 19:   E. Ehlers, C. F. Gethmann (ed) Environmental Across Cultures, 2003

Vol. 20:   R. Chadwick et al., Functional Foods, 2003

Vol. 21:   D. Solter et al., Embryo Research in Pluralistic Europe, 2003

Vol. 22:   M. Decker, M. Ladikas (eds) Bridges between Science, Society and Policy. Technology Assessment – Methods and Impacts, 2004

Vol. 23:   C. Streffer et al., Low Dose Exposures in the Environment. Dose-Effect Relations and Risk-Evaluation, 2004

Vol. 24:   F. Thiele, R. A. Ashcroft, Bioethics in a Small World, 2004

Vol. 25:   H.-R. Duncker, K. Prieß (eds) On the Uniqueness of Humankind, 2005

Vol. 26:   B. v. Maydell, K. Borchardt, K.-D. Henke, R. Leitner, R. Muffels, M. Quante, P.-L. Rauhala, G. Verschraegen, M. Zukowski, Enabling Social Europe, 2006

Vol. 27:   G. Schmid, H. Brune, H. Ernst, A. Grunwald, W. Grünwald, H. Hofmann, H. Krug, P. Janich, M. Mayor, W. Rathgeber, U. Simon, V. Vogel, D. Wyrwa, Nanotechnology. Assessment and Perspectives, 2006

Vol. 28:   M. Kloepfer, B. Griefahn, A. M. Kaniowski, G. Klepper, S. Lingner, G. Steineach, H. B. Weyer, P. Wysk, Leben mit Lärm? Risikobeurteilung und Regulation des Umgebungslärms im Verkehrsbereich, 2006

Vol. 29:   R. Merkel, G. Boer, J. Fegert, T. Galert, D. Hartmann, B. Nuttin, S. Rosahl, Intervening in the Brain. Changing Psyche and Society, 2007

Vol. 31:   G. Hanekamp (ed) Business Ethics of Innovation, 2007

Also the following studies were published by Springer:

Environmental Standards. Combined Exposures and Their Effect on Human Beings and Their Environment, 2003, Translation Vol. 5

Sustainable Development and Innovation in the Energy Sector, 2005, Translation Vol. 18

F. Breyer, W. van den Daele, M. Engelhard, G. Gubernatis, H. Kliemt, C. Kopetzki, H. J. Schlitt, J. Taupitz, Organmangel. Ist der Tod auf der Warteliste unvermeidbar? 2006

CPSIA information can be obtained at www.ICGtesting.com
Printed in the USA
LVOW070413110412

277094LV00004B/1/P

9 783540 723097